D1480293

How to Fall Slower Than Gravity

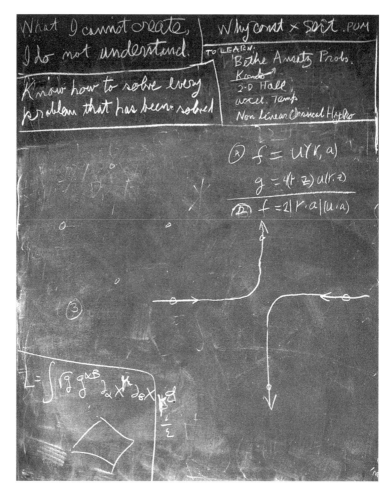

The blackboard in Richard Feynman's Caltech office as he left it when he went home for the last time. Feynman (1918–1988) was one of the great mathematical physicists of the 20th century (he received a share of the 1965 Nobel Prize in Physics for his work in quantum electrodynamics). The photo illustrates the popular view of mathematical physicists: scribes of arcane mystical symbology, understandable by only an elite few. However, I don't think that Feynman himself believed that. As he stated in a famous interview (*Omni*, February 1979), "I don't believe in the idea that there are a few peculiar people capable of understanding math and the rest of the world is normal. Math is a human discovery, and it's no more complicated than humans can understand. I had a calculus book once that said, 'What one fool can do, another can.' [Feynman was referring to the 1910 *Calculus for the Practical Man*, by the British electrical engineer Silvanus P. Thompson.] ... There's a tendency to pomposity in all this, to make it all [artificially] deep and profound." The two principles written in the upper-left-hand corner have become firmly attached to the Feynman legend, and were not just slogans: he vigorously championed them in his writings and talks.

Photo courtesy of the Archives, California Institute of Technology

How to Fall Slower Than Gravity

And Other Everyday (and Not So Everyday) Uses
of Mathematics and Physical Reasoning

PAUL J. NAHIN

PRINCETON UNIVERSITY PRESS
PRINCETON AND OXFORD

Copyright © 2018 by Princeton University Press

Published by Princeton University Press,
41 William Street, Princeton, New Jersey 08540

In the United Kingdom: Princeton University Press,
6 Oxford Street, Woodstock, Oxfordshire OX20 1TR

press.princeton.edu

All Rights Reserved

LCCN 2018936898
ISBN 9780691176918

British Library Cataloging-in-Publication Data is available

Editorial: Vickie Kearn and Lauren Bucca
Production Editorial: Debbie Tegarden
Text Design: Carmina Alvarez
Jacket Design: Anne Karetnikov
Production: Jacquie Poirier
Publicity: Sara Henning-Stout
Copyeditor: Barbara Liguori

This book has been composed in ITC New Baskerville
Printed on acid-free paper. ∞

Typeset by Nova Techset Pvt Ltd, Bangalore, India
Printed in the United States of America

1 3 5 7 9 10 8 6 4 2

For Pat, for Everything

A uniform steel wire in the form of a circular ring is made to revolve in its own plane about its centre of figure. Show that the greatest possible linear velocity is independent both of the [cross]section of the wire and of the radius of the ring, and find roughly this velocity, the breaking strength of the wire being given as 90,000 lbs per square inch, and the weight of a cubic foot [of steel] as 490 lbs.

—A problem posed by John William Strutt (1842–1919), better known in the world of physics as Lord Rayleigh (winner of the 1904 Nobel Prize in Physics), on the fourth day of the famous 9-day (!) Cambridge University Mathematical Tripos Examination of 1876. Lord Rayleigh was the examiner in mathematical physics, and of this question the top scoring man[1] (the so-called Senior Wrangler) remembered it as being "uncommonly high," that is, difficult.

Think about this problem—mentioned in Andrew Warwick's book *Masters of Theory: Cambridge and the Rise of Mathematical Physics*, University of Chicago Press, 2003—as you read, but if after trying really hard you can't solve it (or you just want to check your answer), you'll find an analysis in the final appendix of this book. (Warwick doesn't solve it.)

---

[1] Joseph Timmis Ward (1853–1935), who, with some trepidation, wrote in his diary that same night, "What he [Lord Rayleigh] will give us in the [final] five days I cannot think." Shortly after taking top spot in the Mathematical Tripos, Ward did the same in the even more demanding Smith's Prizes examination. Being good at taking math exams didn't always lead to a successful scientific life; Ward went on to take Holy Orders and, after decades as a priest, died a forgotten recluse in his college rooms at Cambridge. Taking third spot behind Ward in the 1876 Mathematical Tripos, however, was John Henry Poynting (1852–1914), who became famous in physics for the *Poynting vector* (1884), which describes how energy flows in an electromagnetic field.

Contents

PART II. THE SOLUTIONS   155

 *Preface*

> A little mathematics never hurt a practical man. I was
> self-taught and it slowed me down.
> —*Higgston Rainbird, inventor of the first time machine*[1]

One morning about three years or so ago, I sat down with a hot cup of coffee and snapped open that day's *Boston Globe*. After reading the most enlightening section (the comics, of course, with particular attention paid to *Arlo & Janis*), I then turned to the editorial/opinion section. There I found an outrageous—indeed, incredible—letter, a letter that struck me as being far more ludicrous than anything I had just read in the comics. Appearing under the title *Who needs to know this stuff?* the letter was written as a protest against the importance that a particular college placement exam (taken by high school seniors) gave to knowledge of mathematics. The letter contained the following lines:

> It's fun knowing how to solve quadratic equations involving the square root of minus 1. And that knowledge is needed in a few lines of work. But why should it be required for taking entry-level college courses, the gateway to so many good jobs? ...Advanced algebra is the new Latin. It has some uses, but mostly it's a hurdle that trips up young people who could be successful in challenging occupations but don't get the chance because they haven't learned the quadratic formula. Many of those who learn the quadratic formula ...could have used their time more productively learning life skills and important knowledge such as how to reason.[2]

[1] From the short story "Rainbird" by R. A. Lafferty (*Galaxy Science Fiction*, December 1961).

[2] For those who like to check citations, the letter can be found (presumably now only on your local library's microfilm reader) on page A15 of the August 13, 2015, issue of the *Boston Globe*.

Well! What can you possibly say to that, other than it's complete nonsense. Now, the authors of that letter would almost surely reply with "That's just your *opinion*, buddy, and *you* are the one who is wrong." Once I calmed down, I realized that sort of back-and-forth bickering isn't likely to change many minds—and so I didn't write an outraged letter to the editor in reply—but I also realized that a more effective reply *might* be solid demonstrations of the power of mathematical reasoning when applied to the real world, the world the authors of that letter are so rightfully concerned about. And that's what this book—a sequel to my earlier book *In Praise of Simple Physics* (Princeton University Press, 2016)—is about. It's a collection of essays, each not terribly long, each illustrating the ability of mathematics (algebra, trigonometry, geometry, and, occasionally, elementary calculus along with a little help now and then from my laptop computer), in combination with the fundamental physical laws, to provide insight into *real* problems. These problems are not simply textbook calculations designed to give practice in abstract symbol manipulation. In contrast, almost all are strong not only in mathematics but in physical significance, as well. Each essay ends with at least one challenge question for *you* to analyze.

A fair percentage of the essays specifically involve quadratic (and even higher order) equations, and those calculations in particular are why I think the writers of the *Boston Globe* letter were dead wrong. After you read this book, I think you'll agree with me. As I wrote I imagined my audience as readers who took their math and physics courses in high school and undergraduate college years seriously (there are a *million* students alone, worldwide, enrolled in freshman calculus courses each year) but who, for one reason or another, didn't become mathematical physicists. Still, even though today they are, instead, lawyers, biologists, chemists, accountants, computer scientists, engineers, doctors, and poets (maybe even some Hollywood scriptwriters, along with a few professional football players, perhaps), they remember the pleasure of watching the power that mathematical analysis has in making sense of the physical world. That, in fact, is what mathematical physicists *do*. If you are that sort of person, too, then I wrote this book for you on how mathematical physicists *think*.

A survey of the many popular treatments available in bookstores on various topics in mathematics shows that a sizeable fraction of people appreciate the importance of the subject. A similar appearance of mathematics in a different form of popular culture—the movies—sends the same message. For example, *Good Will Hunting* (1997), in which Fourier integrals stream across the opening screen shot; *A Beautiful Mind* (2001), in which allusions to esoteric game theory are sprinkled throughout; *The Man Who Knew Infinity* (2015), which opens with images of the mysterious equations in the 1913 letters the Indian genius Ramanujan (1887–1920) sent to the world-famous English mathematician G. H. Hardy (1877–1947); and *Hidden Figures* (2016), with its many mentions of the mathematics of orbital mechanics. And on television we of course had the major network crime series NUMB3RS (2005–2010) and, most recently, episode 1 of HBO's 2016 series *The Night Of*, which opens with a college student furiously taking notes as his instructor writes on the blackboard and lectures (correctly!) on Stokes's theorem from differential vector calculus.

The distinction between pure mathematics and pure physics is often not a sharp one, and there are some particularly talented people who can work at the highest levels in both worlds. One such person was the Polish-born American mathematical physicist Mark Kac (1914–1984), who, in the introduction to his elegant autobiography, *Enigmas of Chance* (Harper & Row, 1985), wrote with insight on that distinction:

> In mathematics when you discover something you have the feeling that it has always been there. In physics you have the feeling that you are making a real discovery. ... If doing mathematics or science is looked upon as a game, then one might say that in mathematics you compete against yourself or other mathematicians; in physics your adversary is nature and the stakes are higher.

My goal in writing this book has been to provide specific examples of just what Kac was talking about.

Now, you might think it obvious that a *mathematical* physicist cannot function without *mathematics*. Nonetheless, the relationship between mathematics and physics has not always been a smooth one. To quote the Swiss theoretical physicist Res Jost (1918–1990):

> Relations between mathematics and physics vary with time. Right now and for the past few years, harmony reigns and a honeymoon blossoms.

However, I have seen other times, times of divorce and bitter battles, when the sister sciences declared each other as useless—or worse. The following exchange between a famous theoretical physicist and an equally famous mathematician might have been typical, some fifteen or twenty years ago. Says the physicist: "I have no use for mathematics. All the mathematics I ever need, I invent in one week." Answers the mathematician: "You must mean the seven days it took the Lord to create the world."[3]

Well, despite such conflicts, in this book we'll simply assume that mathematics is essential to the work of physicists. But even so, you may well wonder just how mathematical are we going to get in this book. It seems obvious a priori that we have to be at least somewhat mathematical: after all, in a book with the word *mathematical* in the title, wouldn't you feel misled, if not outright cheated, if you *didn't* see equations here and there? Well, *sure*, you agree, that makes sense, *but*, you then ask, will there be lots of *really hard* equations? The short answer is, if you've studied math at a level equivalent to AP calculus, then you are good to go. High school algebra, trigonometry, geometry, and an understanding of what a derivative and an integral are, are all you'll need. Okay, that's the *short* answer. Let me now give you a specific example of the sort of mathematical sophistication you should have.

I have chosen this particular example precisely because it uses only high school level reasoning, and yet I am willing to wager (maybe up to $5!) that even professional mathematicians might have to take a bit of time to do it. The message I hope this example sends is that we are going to be discussing problems that, while requiring only relatively unsophisticated mathematical analysis, will nevertheless result in being

---

[3] From Jost's essay "Mathematics and Physics since 1800: Discord and Sympathy," in *The Fairy Tale about the Ivory Tower: Essays and Lectures* (in German), (ed. K. Hepp, W. Hunziker, and W. Kohn) Springer, 1995. Jost didn't say who the "famous theoretical physicist" and the "equally famous mathematician" were, but in fact the exchange took place during a lecture that Kac gave at Caltech; the physicist (in the audience) was Richard Feynman. Kac and Feynman had known each other since the late 1940s, when they were colleagues at Cornell University.

able to successfully attack nontrivial physics questions. So, here's the problem:

Prove that the product of *any* consecutive m positive integers is *always* divisible by $m!$, that is, by m-factorial, where $m! = m(m-1)(m-2)\ldots(3)(2)(1)$.

This claim says, for example, that $(17)(18)(19)(20)(21)(22)(23)$ is divisible by $7!$, which you can easily confirm by doing the obvious cancellations. But confirming specific examples, no matter how many you do, isn't a proof that the claim is *always* true for *all* possible m. This might seem to be, in fact, a most difficult task, but it isn't. Here's one way to do it.

We start by observing that the claim is trivially true for the $m = 1$ and $m = 2$ cases. The $m = 1$ case says that any single integer is always divisible by $1! = 1$—and I do hope you find that trivial! The $m = 2$ case says that the product of any two consecutive integers (one of which must then be odd and the other even) is divisible by $2! = 2$, and of course that 2 *will* divide the even integer. For $m > 2$, however, matters aren't so obvious anymore.

So, what we'll do is use the method of *induction*, an approach especially beloved by mathematicians. That is, we'll *assume* the claim is true for the case of $m = n - 1$, where n is some integer greater than 2. Then we'll show that from that assumption the claim must be true for the $m = n$ case. Since we already know the claim is true for $m = 2$, then it must be true for $m = 3$, which says in turn that it's true for $m = 4$, and so on, forever. This is the mathematical version of levitating yourself by pulling up on your shoestrings. Physicists can't do that, but mathematicians can! We start by defining the function $\phi_n(r)$ to be the product of n consecutive positive integers, starting with r. That is,

$$\phi_n(r) = r(r+1)(r+2)\cdots(r+n-1).$$

In particular,

$$\phi_n(1) = 1(2)(3)\cdots(n) = n!.$$

Since

$$\phi_n(r+1) = (r+1)(r+2)\cdots(r+n-1)(r+n),$$

then

$$\phi_n(r+1) - \phi_n(r) = [(r+1)(r+2)\cdots(r+n-1)][(r+n)-r],$$

or

$$\phi_n(r+1) - \phi_n(r) = [(r+1)(r+2)\cdots(r+n-1)]n.$$

The quantity on the right, in the square brackets, is the product of $n-1$ consecutive integers and so, *by our assumption*, is divisible by $(n-1)!$ That is,

$$[(r+1)(r+2)\cdots(r+n-1)] = \text{multiple of } (n-1)!,$$

and so

$$\phi_n(r+1) - \phi_n(r) = \{\text{multiple of } (n-1)!\}\, n = \text{multiple of } n!,$$

which says

$$\boxed{\phi_n(r+1) = \phi_n(r) + \text{multiple of } n!.}$$

For $r = 1$, this result says

$$\phi_n(2) = \phi_n(1) + \text{multiple of } n!$$

but, as $\phi_n(1)$ is itself equal to $n!$ (take a look back at the second box), then

$$\phi_n(2) = \text{multiple of } n!$$

But then

$$\phi_n(3) = \phi_n(2) + \text{multiple of } n!$$
$$= \text{multiple of } n! + \text{multiple of } n! = \text{multiple of } n!$$

and so on for $\phi_n(4)$, $\phi_n(5)$, ... and so on, forever! That is, for r any positive integer,

$$\phi_n(r) = \text{multiple of } n!$$

and we are done. What a clever bit of analysis, don't you think?[4]

A physicist can appreciate such a pretty demonstration as well as any mathematician can, but he or she can also resist the temptation to be swept up by the gloriousness of it all. In his autobiography, for example, Kac tells a revealing story, about a comment made by a physicist to a mathematician at the famous MIT Radiation Laboratory:

"You can keep your Hilbert space," said the physicist, "I want the answer in volts."[5] That'll be the tone of the rest of this book.

Kac, who spent nearly his entire career in academia, had an international reputation as a first-rate probability analyst, but during World War II he also worked part-time at the Radiation Laboratory on the real-life physical problems caused by electronic noise in radar. Being blessed with such ambidextrous talent is not necessarily appreciated by those whose abilities are restricted to just one field or the other. For example, when Kac published a paper (in the *Journal of Applied Physics*) after the war, based on his radar work, he received a postcard from a friend—the brilliant Hungarian pure mathematician Paul Erdös (1913–1996)—with this lone sentence: "I am praying for your soul."

To maximize your pleasure, the book is structured as a sequence of problem statements, with all their solutions gathered together at the end. That way you can try your hand at each problem, on your

---

[4]This is not self-praise. I came across this derivation years ago while reading a 19th-century textbook by the English mathematician W. W. Rouse Ball (1850–1925): *Elementary Algebra*, Cambridge University Press, 1890, p. 415. Whether this was his original discovery or if he in turn found it in an even earlier work, I don't know.

[5]The physicist was Samuel Goudsmit (1902–1978), who is famous as the chief scientific advisor of the U.S. Alsos Mission (immediately after the June 1944 Normandy invasion of Europe) to uncover the details of the Nazi chemical, biological, rocket, and atomic weapons programs, and the mathematician was the famously eccentric Norbert Wiener (1894–1964).

own, before reading my solution, but if you get stuck, or just want to compare your approach with mine, you can then turn to the solutions section. (One physicist who looked at an early draft of this book thought readers would "just" read the problems and then immediately turn to the solutions, to which I replied "And that's bad?" If that's what *you* do, that's okay, too. After all, it's *your* book!) To show you how that will work, consider the following math challenge problem.

Challenge Problem 1: The Hellenized Babylonian mathematician Diophantus of Alexandria—who is thought to have lived around AD 250, six centuries after Euclid—is remembered[6] today as the originator of the following class of problems: given a single polynomial equation (called a *Diophantine equation*) in n variables, do there exist any *positive integer* solutions? For example, what are all the positive integer values (if any) for x and y such that

$$72x + 694y = 1,001,001?$$

Similarly, what are the positive integer values (if any) for x, y, and z such that

$$x^3 + 2y^3 = 4z^3?$$

To answer these two questions (the first, despite the big coefficients, is easy, and the second, despite the small coefficients, is not so easy) requires only the elementary concepts of evenness and oddness. If you get stuck, solutions to both questions are given in the very first entry of the solutions section.

---

[6] Little is known of Diophantus's life, other than how old he was when he died. That's because of a puzzle, dating to the 4th century, that has probably appeared in every introductory algebra text written since: "His boyhood lasted 1/6 of his life; his beard grew after 1/12 more; after 1/7 more he married, and his son was born 5 years later; the son lived to half his father's age, and the father died 4 years after his son." So, if Diophantus died at age x, we have

$$\frac{1}{6}x + \frac{1}{12}x + \frac{1}{7}x + 5 + \frac{1}{2}x + 4 = x,$$

and I'll let you solve for x.

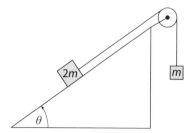

Figure 1. The two masses move at a constant speed.

Challenge Problem 2: This type of problem often comes up during a theoretical analysis. Using *only algebra* (no calculus!) derive the power series expansion of $\sqrt{1+x} = c_0 + c_1 x + c_2 x^2 + c_3 x^3 + c_4 x^4 + \cdots$. That is, what are the c's? You can look up this particular expansion in a math handbook, but a mathematical physicist should know how to do it from scratch. What if, for example, the answer is key to escaping from a remote island, and all you have is a stick with which to write in the sandy beach? (Don't laugh, that's at least a *conceivable* situation!) The solution is the second entry in the solutions section.

And since this is a book on mathematical *physics*, here's a third, *physics* challenge problem for you to consider.

Challenge Problem 3: In Figure 1 a mass $2m$, on an incline of angle θ, is connected via a string and pulley to a hanging mass m. The constant coefficient of kinetic friction between the mass on the incline and the incline is μ.[7] The two masses are observed to move at a constant speed. (This is a clue to think of Newton's second law of motion, the famous "force is mass times *acceleration*.") What is the relationship between θ and μ? Specifically, plot that relationship (pay special attention to the interval $0 \le \mu < \frac{1}{2}$), and comment on anything you find particularly interesting. An analysis is the third entry in the solutions section.

---

[7]Recall that if a moving mass applies a force F_n (say, its weight) *normal* to the surface on which it is moving, and if the mass experiences a frictional *drag* force (that is, a force *opposite* the direction of motion) *along* the surface of F_r, then $\mu = \frac{F_r}{F_n} \ge 0$ is called the *kinetic coefficient of friction*.

While the analytical techniques used in this book will, for the most part, be straightforward, occasionally we'll get a little tricky, too. To illustrate what I mean by a trick, consider the following little puzzle that I recall first hearing in high school. An old farmer died and, in his will, left all the cows in his barn to his three sons, Al, Bob, and Chuck. Al was to receive one-half of the cows in the barn, Bob was to receive one-third of the cows in the barn, and Chuck was to receive one-twelfth of the cows in the barn. The problem was that when the farmer died there were 11 cows in the barn! Everybody was at a standstill on how to carry out the farmer's last wishes, until the farmer's brother (a retired mathematical physicist who owned a few cows himself) solved the problem. Here's what he did.

(1) He placed one of his own cows in the barn, and so now there were 12 cows in the barn;
(2) Al got 6 cows (1/2 of the 12 cows in the barn), leaving 6 cows in the barn;
(3) Bob got 4 cows (1/3 of the 12 cows in the barn), leaving 2 cows in the barn;
(4) Chuck got 1 cow (1/12 of the 12 cows in the barn), leaving 1 cow in the barn;
(5) The retired mathematical physicist took the one cow left in the barn—the very cow, in fact, that he placed in the barn in step (1)—and went home.

Notice that even if the mathematical physicist hadn't actually had any cows of his own, he could have simply put an imaginary cow in the barn and then, in step (5), have taken his imaginary cow back. Now *that's* a trick, and we won't be above doing stuff like that now and then![8] Isn't math wonderful?

Just so you won't think I'm going to be putting too much emphasis on tricks, let me show you one more "serious" example of the

---

[8]As mathematicians like to say, a trick you can use more than once is a *method*. At the end of this preface I'll give you two more challenge problems, both of ancient origin dating to the first century AD in the Chinese mathematical literature, both of which can be solved by using the "imaginary cow" *method*. Mathematical physicists will enjoy these problems, too.

usefulness of the quadratic formula to mathematical physicists. It is amazing how often, when reading a technical paper in a physics journal, one runs into a sentence reading something like this: "Now, if we apply the Cauchy-Schwarz inequality[9] to our last result we immediately see that . . . " The Cauchy-Schwarz inequality says

$$\left\{ \int_a^b f(x)g(x)dx \right\}^2 \le \left\{ \int_a^b f^2(x)dx \right\} \left\{ \int_a^b g^2(x)dx \right\},$$

where $f(x)$ and $g(x)$ can be just about any real-valued functions you wish, just as long as the definite integrals exist. This is a general theorem of astonishing value to mathematical physicists; one mathematician called it "an exceptionally potent weapon."[10] The same mathematician then went on to correctly write "There are many occasions on which persons who know and think of using this formula can shine while their less fortunate brethren flounder." Later in this book I'll invoke the Cauchy-Schwarz inequality in one of the problem discussions, but *how* (you may be wondering right now) do we *prove* it? Well, all it takes is the quadratic formula. Here's how it goes.

With λ as an arbitrary parameter, it is certainly true that

$$\int_a^b \{\lambda f(x) + g(x)\}^2 \, dx \ge 0,$$

because the integral of any real function *squared* can't be negative (again, think of the area interpretation of the definite integral). So, expanding this integral out in detail, we get

$$\lambda^2 \int_a^b f^2(x)dx + 2\lambda \int_a^b f(x)g(x)dx + \int_a^b g^2(x)dx \ge 0.$$

---

[9]Named after the French mathematician Augustin-Louis Cauchy (1789–1857) and the German mathematician Hermann Schwarz (1843–1921). To understand the physical significance of this inequality, all you really need to know is that the definite integral of a function $h(x)$, over the interval $a \le x \le b$, is the *area* under the curve $y = h(x)$ as x varies from a to b.

[10]Ralph Palmer Agnew, *Differential Equations*, McGraw-Hill, 1960, p. 370.

All these integrals, as *definite* integrals, have *definite* values:

$$\int_a^b f^2(x)dx = A, \quad \int_a^b f(x)g(x)dx = B, \quad \int_a^b g^2(x)dx = C.$$

That is,

$$A\lambda^2 + 2B\lambda + C \geq 0,$$

a quadratic inequality in λ. The $\geq$ sign means, "physically," that a plot of the left-hand side, as a function of λ, *never crosses* the λ-axis (think of the λ-axis as the horizontal axis). That is, the left-hand side is a curve that is always *above* (or, at most, *just touching*[11]) the λ-axis. Not crossing the λ-axis means there are no *real* solutions (the crossing points *are* the solutions) for λ to

$$A\lambda^2 + 2B\lambda + C = 0.$$

Since the quadratic formula tells us that the solutions are

$$\lambda = \frac{-2B \pm \sqrt{4B^2 - 4AC}}{2A} = \frac{-B \pm \sqrt{B^2 - AC}}{A},$$

then the condition for no *real* solutions is

$$B^2 - AC < 0,$$

which of course gives *complex* solutions. That is,

$$B^2 < AC,$$

or

$$\left\{\int_a^b f(x)g(x)dx\right\}^2 < \left\{\int_a^b f^2(x)dx\right\}\left\{\int_a^b g^2(x)dx\right\},$$

[11]*Just touching* happens, of course, when the $\geq$ sign becomes the special case of equality.

which *is* the Cauchy-Schwarz inequality.[12] That's it! The authors of the *Boston Globe* letter are now (I hope) enthusiastically *eager* to reconsider the error of their argument!

As I wrote earlier, all the specific examples in this preface have been chosen with the goal of illustrating "how physicists think." Now, that's a pretty bold claim to make, and so let me end with yet three more examples to make my case. First, in observing the real world, it seems to be the nature of Nature that there really isn't such a thing as a *discontinuous force*.[13] The idea of *continuity* in physics is a powerful one, and it can be used to easily establish results that would otherwise be far more difficult to achieve. As an example, imagine a man who is about to walk up a hill, starting at A (the bottom of the hill) and ending at B (the top of the hill). You know nothing of *how* he walks (perhaps at times he stops for a while, other times he walks slowly, sometimes he walks briskly, and perhaps at times he even walks back down the hill). All you know is that, starting at A at 10 o'clock in the morning, he arrives at B at 11 o'clock. That is, the walk up the hill takes exactly 60 minutes.

He camps overnight at B and then, the next morning at exactly 10 o'clock, he walks back down the hill, along the same path he followed during his ascent the previous day. He arrives at A at 11 o'clock. That is, the return trip takes exactly 60 minutes. Again, you know nothing of *how* he makes the descent.

When told all this, a mathematical physicist can *immediately* draw the following conclusion: there is at least one spot on the path traveled that the man passed at exactly the same time during his descent as the

---

[12]The *just touching* condition says $B^2 = AC$ (giving a real, *double* root to the λ quadratic), and that's why the $<$ sign in the Cauchy-Schwarz inequality is replaced with a $\leq$ sign.

[13]When I write that, I am actually copying two well-known mathematical physicists: Richard Feynman, and his Princeton PhD dissertation advisor, John Wheeler (1911–2008), who invoked continuity of force in a famous 1949 paper dealing with a particularly puzzling paradox of time travel (see my book *Time Machine Tales: The Science Fiction Adventures and Philosophical Puzzles of Time Travel*, Springer, 2017, pp. 264–269). You can find more about these two very important figures in 20th-century physics in Paul Halpern's book *The Quantum Labyrinth: How Richard Feynman and John Wheeler Revolutionized Time and Reality*, Basic Books, 2017.

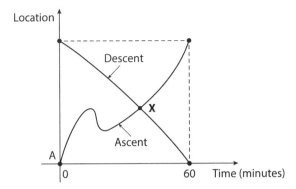

Figure 2. X marks the spot, same place, same time.

time at which he passed that same spot during the ascent. The physicist can make this claim without writing a single equation but rather needs only to invoke continuity. Here's how. In Figure 2 we see two plots, each showing the location (as measured from A along the path) of the man versus time. One plot is for the ascent, and the other is for the descent. Each plot is *continuous*, and there are no other conditions.[14] There is *no way* you can draw those two plots without there being at least one intersection point (the big **X**), and such a point is the same distance from A at the same time for one plot as for the other plot.

Not a single equation, anywhere—Just a lone sketch and a physical argument.

We can do even this almost minimal analysis one better, however, with an initially puzzling (for most people) problem that can be done with nothing but prose (and a clever physical argument). Mark n dots on a sheet of paper, where n is any positive integer you wish (as long as $n \geq 2$). Then, as many times as you wish, select any two dots and connect them with a line. Any particular dot may be chosen multiple times or, perhaps, not at all. At all times, each dot will appear to be radiating lines. When you get tired of selecting pairs of dots and drawing lines, stop. The dots can then be divided into two sets: the set of dots that radiate an even number of lines, and the set of dots that

---

[14]Other than each plot must never double back along the time axis. Wheeler and Feynman used a continuity argument in a time-travel setting, but we are being just a bit more conservative here!

radiate an odd number of lines. Prove that this second set will always have an even number of dots in it.

Where does one start with something like this? Well, I think the following analysis is a beautiful illustration of how a mathematical physicist would approach such a question. Imagine that next to each of the n dots is a counter displaying how many lines are currently radiating from the dot. So, before any lines are drawn, all the counters read zero, and the *total* count for all the dots is zero. Now, draw a line connecting any two dots. The counters for those two dots are each incremented by 1, and the total count for all the dots is 2. Draw another line. The counters for those two dots each increment by 1, and so the total count for all the dots is 4, and so on, with the result that the *total* count for all the dots is always an *even* number. When the drawing of lines is finished, the counter for each individual dot will display either an odd number or an even number. The sum of the displays for those dots with even counters will of course be even (the sum of any number of even numbers is even). Thus, the sum of the displays for those dots with odd counters will also be even (that sum is the total sum over all dots—which is even—minus the sum over the even counters, which we just argued is also even), and of course even minus even is even. So, we have a number (call it N) of dots, each with a counter displaying an odd number, with the sum of those displays being even. So, N times odd $=$ even and therefore N must be even because the alternative doesn't work (*odd* times odd is odd).

Who could argue against that being anything other than just pretty darn slick?

As a final example of the physical way of thinking about mathematics (but now with some actual mathematics involved), consider the so-called mean value theorem, a topic discussed in every freshman calculus class. Suppose that the function $f(y)$ is continuous over the interval $a \leq y \leq b$ and that it has a derivative everywhere in that interval. That is, $f(y)$ represents what mathematicians call a *smooth curve*. Suppose further that we take any two points in that interval, say y_1 and y_2, and draw the chord joining $f(y_1)$ and $f(y_2)$, as shown in Figure 3.

The mean value theorem states that there exists a specific value of y—call it y^*—such that the slope of the chord is equal to $f'(y^*)$—the

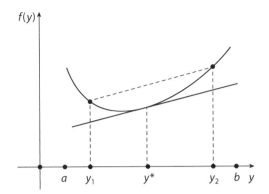

Figure 3. A chord (dashed), and a tangent line (solid) of a curve.

slope of the tangent line to the curve at $y = y^*$, where f' is the derivative of $f(y)$. The theorem doesn't say *where* y^* is, only that it exists. This is an example of a mathematical statement that a mathematical physicist would declare to be "obvious." It has, however, implications that are far from being trivial.

Consider, for example, its use in deriving the decidedly nonobvious *logarithmic inequality*, which we'll use later in this book. If we define $f(y)$ to be the natural log function, that is,

$$f(y) = \ln(y),$$

then

$$f'(y) = \frac{1}{y}.$$

A plot of $f(y)$ is shown Figure 4, where we consider the interval $1 \le y \le 1 + x$, for $x \ge 0$.

The mean value theorem says there is some $y = y^*$ in the interval 1 to $1 + x$ such that

$$\frac{1}{y^*} = \frac{\ln(1+x) - \ln(1)}{(1+x) - 1} = \frac{\ln(1+x)}{x}.$$

That is,

$$\ln(1+x) = \frac{x}{y^*},$$

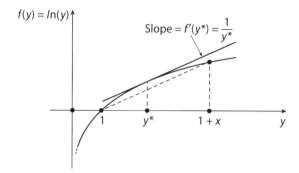

Figure 4. The natural log function, with a chord and a tangent line.

where y^* is *somewhere* (remember, we don't know where) in the interval 1 to $1+x$. If we replace y^* (whatever it is) with a *larger* quantity (such as the right end of the interval, $1+x$), then clearly the left-hand side of the preceding equality becomes $\geq$, while if we replace y^* (whatever it is) with a *smaller* quantity (such as the left end of the interval, 1), then clearly the left-hand side of the equality becomes $\leq$. So, just like that, we have

$$\ln(1+x) \geq \frac{x}{1+x},$$

and

$$\ln(1+x) \leq x,$$

or, writing the equations as a double inequality,

$$\frac{x}{1+x} \leq \ln(1+x) \leq x, \quad x \geq 0.$$

This result[15] will be central to the discussion in Problem 15.

Finally, to end this preface, let me say something about the computer codes—included to illustrate algorithms that solve problems we can't

---

[15] It is easy to extend the validity of the double inequality to $x > -1$, and I'll leave doing that for your pleasure.

do analytically—that are sprinkled throughout the book. They are all written in MATLAB, a programming language commonly taught to engineering students. I use MATLAB because I am a retired electrical engineering professor, and MathWorks (the creator of MATLAB) has kindly given me a license to run MATLAB for free on my personal home computer. Well, you say, "How nice for *you*, Professor Nahin, but *I* don't have MATLAB—what am *I* supposed to do?"

Here's my answer: Study the algorithm descriptions that are given in the book, and then write *your own* codes in the language of your choice. Mathematicians seem to be particularly fond of Mathematica, while physicists appear to be equally split between MATLAB and Mathematica. (One early reviewer of this book made a pitch for Python, which has the advantage of being available with an open-source license from MIT.) Just remember: a computer language is simply a tool. If your computer runs only BASIC, don't worry about it. What counts is the *answer*. All my MATLAB codes are written in a *very* low-level way, using only well-known, common commands (IF, WHILE, FOR, and so on), and so should be easy to convert to just about any other language. To give you a little taste of the power of computers (and of MATLAB and other programming languages like MATLAB) in the study of mathematical questions, I've included a short "primer by example" at the end of the book (immediately after the solutions) that you can read at any time.

The digital computer has long since proven itself to be an indispensable tool in the study of physical systems whose behavior would otherwise be unobservable. This lack of observability, for example, might be the result of such practical limitations as an insufficient supply of money and/or time (for instance, neutron diffusion in a new and expensive nuclear reactor wall design, or the evolution of galactic structures under Newtonian mechanics over vast spans of time). With the relevant physics programmed into a computer, however, we don't actually have to build the reactor wall to determine what the leakage flux will be, and even more dramatically, we can be almost godlike and run the universe in *faster* than real time (or even *backward* in time).

Even more interesting, perhaps, are those problems in which the *fundamental physics* is altered from what we take it to be in the real world (the nature of force interaction and the conservation laws can be modified as easily as one can type new lines of computer code). This last possibility, involving what some might call *impossible physics*, I will refer to here as representing *imaginary* systems, and the application of computers to many such problems is well known to physicists. In Problem 17, for example, you'll find an example of that with the simulation of the mysterious NASTYGLASS.

While perhaps easy to dismiss as simply another tool to be used by physicists, engineers, and mathematicians, as we might say of pencils, pens, and paper, I think computers and programming languages are far more than that. The impact of computers on our daily lives has exponentially increased, and surely that explosive growth will continue: mathematical physics will be only one of countless areas of human experience to feel the significance of that growth.

Three More Challenges
(Solutions at the End of This Book)

The first two of these challenge problems—Challenges 4 and 5—have been taken from the classic college text *Elementary Number Theory*, by David M. Burton.[16] Professor Burton provides some of the numerical answers at the end of his book but gives no indication of how to *derive* those numbers. That's *your* task here. Don't launch a head-on, brute-force, sledgehammer attack, but take a few moments to look for a trick (that is, a *pivotal insight*). *Hint*: Think of imaginary cows!

Challenge Problem 4: When eggs in a basket are removed 2, 3, 4, 5, 6 at a time there remain, respectively, 1, 2, 3, 4, 5 eggs. When they are taken out 7 at a time, none are left. Find the first three numbers of eggs that could be in the basket.

---

[16] Professor Burton (1930–2016), late professor emeritus of mathematics at the University of New Hampshire, was a colleague and friend for decades. We spent many hours together over the years discussing the challenges of writing on the history of mathematics, and it is my real pleasure to both reproduce these two problems from his book and to recommend *Elementary Number Theory* (McGraw-Hill, in numerous editions) to all who love mathematics.

Challenge Problem 5: The basket-of-eggs problem is often cast in the following alternative form: One egg remains when the eggs are removed from the basket 2, 3, 4, 5, or 6 at a time; but no eggs remain if they are removed 7 at a time. Find the first three numbers of eggs that could be in the basket.

Challenge Problem 6: Here's a challenge question that connects the mathematics and some interesting "physics" of drawing a straight line in an infinite plane. Each of the infinity of points in the plane that have integer coordinates is called a *lattice point*. For example, $(0, 0)$, $-3, 2)$, and $(91,674)$ are lattice points. The point $(\pi, \sqrt{2})$ is *not* a lattice point. Prove that an infinitely long straight line, drawn at random, passes through *exactly* zero lattice points, through *exactly* one lattice point, or though an *infinity* of lattice points. (It is *impossible* to draw a straight line that *exactly* passes through any finite number of lattice points other than 0 or 1.) High school math is all that is required (and not much of that, either).

Challenge Problem 7: Finally, to wrap up this preface, here's a *physics* challenge that can be answered without writing a single equation. Imagine that you are standing on the Moon's surface with a hand extended straight out from your body, palm up. In your palm is a golf ball. With a quick flip of the wrist you send the ball upward with some initial speed. The ball rises, slowing down all the while (because of gravity), and after a time t_u reaches some maximum height. It then begins to descend at ever-increasing speed (because of gravity) and eventually falls back into your hand after having fallen for the additional time t_d. Since there is no air (drag) on the Moon, $t_u = t_d$. Now, suppose you do this same experiment on Earth, where there *is* air (drag). Explain why $t_u \neq t_d$. Indeed, explain why $t_u < t_d$. You do *not* need to know any of the detailed physics of air drag (about its dependency, for example, on the instantaneous speed of the ball). *Hint*: Think about conservation of energy, and why $t_u = t_d$ *on the Moon*.

Let me close this preface with these words from Richard Feynman, from a verse made up by him (in an assigned essay on introspection) for a philosophy class he took when an undergraduate, in the late 1930s, at MIT:

> I wonder why. I wonder why.
> I wonder why I wonder.
> I wonder *why* I wonder why
> I wonder why I wonder!

Feynman tells his readers how he came to be motivated to write that, as a sort of a semi-serious joke, in his 1985 book *Surely You're Joking, Mr. Feynman!* But the great Polish-born American mathematician Stanislaw Ulam (1909–1984) recalled those words, in his 1976 autobiography, *Adventures of a Mathematician*, as being part of a quite serious conversation he had with Feynman when both were at Los Alamos during the U.S. World War II atomic bomb program. At one point Feynman recited that same verse to Ulam in such a way as to make a deep impression on the mathematician, and for good reason. They are words that aptly describe the ever-curious nature of the mathematical physicist's mind. I hope that what you read here, if I have been successful, will get you to "wondering why," too.

Paul J. Nahin
Exeter, New Hampshire

In support of the theoretical calculations performed in this book, software packages developed by The MathWorks, Inc. of Natick, MA, were used (specifically, MATLAB 8.1 Release 2013a and Symbolic Math Toolbox 5.10), running on a Windows 7 PC. This software is now several releases old, but all the commands used in this book work with the newer versions, and are likely to continue to work for newer versions for several years more. The MathWorks, Inc. does not warrant the accuracy of the text of this book. This book's use or discussions of MATLAB does not constitute an endorsement or sponsorship by The MathWorks, Inc. of a particular pedagogical approach or particular use of MATLAB, or of the Symbolic Math Toolbox software.

How to Fall Slower Than Gravity

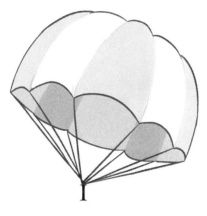

THE PROBLEMS

PROBLEM 1
A Military Question: Catapult Warfare

Our opening problem has a military flavor. Imagine an invading army faced with a huge, VERY tall defensive wall. (Think of The Wall in HBO's *Game of Thrones*, protected by the Night Watch.) To breach the wall, the invaders decide to attack by launching massive projectiles, to hit as high up the face of the wall as possible. (For a particularly nasty, *really disgusting* attack, one not all that uncommon in ancient times, think of a catapult flinging wooden barrels full of fresh cow dung and/or the dead bodies of animals and soldiers up onto—even better over—the wall.) Given that the launching device (catapult, cannon, whatever) gives each projectile a "muzzle" speed of V, the launcher is distance D from the base of the wall, and g is the acceleration of gravity, you are to calculate the launch angle θ that maximizes the height h of the projectile's impact point on the wall (see Figure P1.1). Indeed, what *is* this maximum height? Additionally, what's the flight time, from launch to impact, of the projectile when h is maximized? In all your calculations, ignore the effects of air resistance. *Note*: This problem can be done with nothing but algebra, a touch of trigonometry/geometry, and the quadratic equation. No calculus is required (other than knowing that distance is the integral of speed). No derivatives are required.

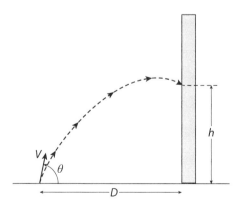

Figure P1.1. What is θ to maximize h, given V and D?

PROBLEM 2

A Seemingly Impossible Question: A Shocking Snow Conundrum

Before I tell you about our second problem—which will also involve a quadratic equation—let me first tell you a little mathematical joke that you may recall having heard in high school. A bear wakes up one morning and, feeling a bit hungry, decides to see what might be available for breakfast. He ambles 1 mile south and then, seeing nothing of interest, changes direction and trots 1 mile west. Again spotting nothing tasty, he decides to again try a new direction and travels a 1 mile north. Striking out yet again on his quest, and thinking he *is* getting a bit fat (as well as noticing he is now back at his comfy home), he grumbles "To heck with it" to himself and goes back to bed. What color is the bear?

When most students first encounter this tale there is generally a shocked reaction—*What color is the bear?*—because the question seems such a non sequitur. In fact, it isn't, and the whole business is, I think, a great way to introduce students to how geometry on a sphere is different from geometry on a plane. The answer is, of course, that the bear is white, because there is only one place on the spherical Earth where the bear's bed could be (the North Pole, where there are only polar bears). Starting at the North Pole, *every* direction is south, and then one mile west and one mile north gets the bear back to the North Pole and into bed.

There is an amusing final twist to this tale that is often missed—the North Pole actually *isn't* the only point on Earth where the bear's walk could start. Such a walk could take place near the South Pole, too (but of course there are no bears of *any* color there, so "the bear is white" is indeed the correct answer). To see this other possibility for the *walk*, however, take a look at Figure P2.1, which shows the North Pole walk at the top and the alternative walk at the bottom. The way the bottom walk works is as follows.

The little circle just above the South Pole is such that its circumference is 1 mile. So, starting at any point on a larger circle that is 1 mile above the little circle, you walk 1 mile south down to the little circle, then 1 mile west around the little circle, which of course brings you

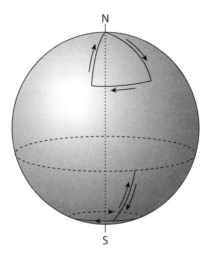

Figure P2.1. The geometry of 'the walk' (not to scale!)

back to your starting point on the little circle, and so the final part of the walk (1 mile north) brings you right back to where you began. In fact, this idea works if the little circle has a circumference of $\frac{1}{n}$ miles, where n is any positive integer. One just goes around the little circle n times. There is an infinity of such little circles, and for each one, there is a larger circle 1 mile north. So, there are *lots* of places near the South Pole to take the walk, but only at the North Pole are there bears available to "walk the walk."

Okay, that's a fun math problem, but it is left in the dust by the following *mathematical physics* problem that you are to consider here. One day in the morning it starts to snow at a steady rate. At precisely noon, a snowplow starts to clear a long, straight road. The plow removes snow at a constant rate (some fixed volume of snow per hour). The driver, an observant fellow, notices that during the second hour of clearing he travels exactly half the distance he traveled the first hour. When did it start to snow?

Again, I think, shock! The problem statement appears to tell us so little; there is no information about how much snow falls per hour, no information on how wide the plow blade is or the speed of the plow, and no information on how far the plow travels the first hour. Can this problem actually be done? The answer is *yes*, and you can calculate *to the very second* when the snow began to fall. You will need to do a bit of freshman calculus, as well as solve a quadratic equation, but it is, despite initial appearances, a well-defined problem. It is, I think, a wonderful example of how mathematical reasoning, and some judicious physical assumptions, can illuminate a path through a blinding blizzard of *seeming* ambiguity.

PROBLEM 3

**Two Math Problems: Algebra and
Differential Equations Save the Day**

Now, for a change of pace. Mathematical physicists, you won't be surprised to learn, are generally pretty good mathematicians (or, at least, they are good *applied* mathematicians). Here are two freshman-level math problems (a high school honors AP Calculus student may perhaps be able to have some success here, too) that are the sort of calculations mathematical physicists often encounter. The first requires only algebra, while the second does require some basic calculus.

(1) Suppose a, b, and c are the lengths of the sides of some (any) triangle. (The word *triangle* is important here.) Show that $abc \geq (a + b - c)(b + c - a)(c + a - b)$. *Hint*: The crucial observation here is the physical fact that the shortest distance between two points in a plane is a straight line.

(2) Most mathematical physicists can't get through even a single day without doing at least one integration. Here's a pretty little integral that looks so simple that at first glance you might think it almost trivial, but a little thought will show you that it does have a bit of a twist to it that is not usually seen in a freshman calculus class. To start, let me make two definitions, just to be sure you understand the notation. For x any number, let's write

$$[x] = \text{integer part of } x,$$

and

$$\{x\} = \text{fractional part of } x.$$

For example, if $x = 3.6173$, then $[x] = 3$ and $\{x\} = 0.6173$. The connection between $[x]$ and $\{x\}$ is easy to write as, obviously, $x = [x] + \{x\}$, and so $\{x\} = x - [x]$. With all this in mind, evaluate

$$\int_0^1 x \left\{ \frac{1}{x} \right\} dx = ?$$

Hint: There is an exact expression for the answer, with the numerical value of $0.17753\ldots$.

You'll find it helpful to know that $\sum_{k=1}^{\infty} \frac{1}{k^2} = \frac{\pi^2}{6}$, a result due to the great Swiss-born mathematical physicist Leonhard Euler (1707–1783), whose derivation of it in 1734 made him famous in the world of mathematics. Now, if you're feeling *really* mathematically powerful, you should try your hand with the following:

Extra Credit Challenge: An obvious extension of our original integral is the evaluation of $\int_0^1 x^2 \left\{ \frac{1}{x} \right\}^2 dx$. This will lead you not only to $\sum_{k=1}^{\infty} \frac{1}{k^2}$, the sum Euler did, but also to $\sum_{k=1}^{\infty} \frac{1}{k^3}$, a sum Euler could *not* do (and nobody else has been able to do, either, to this day, other than to evaluate it numerically as $1.2020569\ldots$). The general sum $\sum_{k=1}^{\infty} \frac{1}{k^n}$ with $n > 1$ is written by mathematicians as $\zeta(n)$, and is called the *zeta function*. Euler was able to calculate exact expressions for $\zeta(n)$ when n is any *even* positive integer—in Problem 24 I'll show you how to do it the modern way, using mathematics that all physics, math, and engineering students see by the end of their second year of college—but $\zeta(n)$ for all the *odd* positive integer values of $n > 1$ has, now for the nearly three centuries after Euler, completely stumped the world's best mathematicians. (For $n = 1$, it was known for centuries *before* Euler that the zeta function blows up, that is, sums to infinity.) So, when you evaluate $\int_0^1 x^2 \left\{ \frac{1}{x} \right\}^2 dx$, leave your answer in terms of $\zeta(2)$ and $\zeta(3)$. *Hint*: The integral's numerical value is 0.051. Another generalization of $\int_0^1 x \left\{ \frac{1}{x} \right\} dx$ is $\int_0^1 x^n \left\{ \frac{1}{x} \right\} dx$ for n *any* positive integer. See if you can do this more general integral (your answer should, for $n = 1$, reduce to your first result).

PROBLEM 4
An Escape Problem: Dodge the Truck

Here's a pretty little problem that, besides involving a quadratic yet again, offers you a way to maximize your chances of surviving a dangerous situation. Imagine that you're jogging down the middle of a straight city street, at your top speed V_y. Suddenly, you become aware of a wide-body truck coming up behind you at speed $V_t > V_y$. If you don't get out of the way, you're going to be run over! So, when the truck is distance S behind you, you decide you'd better get off the street, and to do that, you keep running at full speed but now at an angle θ to your original path (that is, *into* the page), as shown in Figure P4.1.

What should θ be to maximize your distance from the center of the street? And for that best θ, what is the distance you'll be from the center of the street as the truck passes by you? If that distance is greater than half the width of the truck, then you survive. (*Note:* Picking the angle $\theta = 90°$, the angle you might at first choose because it moves you directly away from the street center, is not the proper choice because while $\theta < 90°$ doesn't move you *directly* away from the street center, it moves you *some*, as well as keeps you moving, too, in the direction directly *down* the street and away from the truck.) Now, in particular, if you're running at 15 miles per hour (that's the speed of a 4-minute mile, and so you're a pretty good jogger!), and the truck is traveling at 60 miles per hour, and you begin your escape attempt when the truck (which is 8 feet wide) is 75 feet behind you, do you survive? Does your answer change if you jog at just 4 mph?

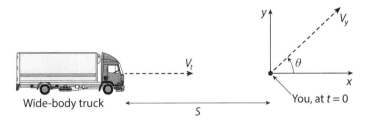

Figure P4.1. Mathematical physics to the rescue!

PROBLEM 5

The Catapult Again: Where Dead Cows *Can't* Go

Here's a variation on the first problem, in which we had a catapult tossing a projectile at a wall. In this version, forget about the wall for the time being. We start with a catapult, at $x = 0$, that flings/fires/tosses the projectile at an initial speed V, at angle θ. There is some launch angle at which the range of the projectile is maximum (you probably already know that the angle for maximum range is 45°, but if not, it's easy to show). Let's call this maximum range R, which is easy to express in terms of just V and the acceleration of gravity, g. Now, introduce the wall, of height H. The catapult is to the left of the wall, distance D away. That is, the wall is at $x = D$. The projectile is to pass over the wall and land on the other side, and you are to find where it is *impossible* for the projectile to land. An aside: The concept of "tossing stuff" didn't disappear with the development of modern weapons. The Air Force embraced tossing, for example, when it was confronted with the task of delivering atomic bombs and the problem of escaping the blast of its own device. One solution was *toss bombing*. As the aircraft, shown in Figure P5.1, approached a ground target, it would pull up into a vertical climb and turn to reverse direction. As it was partway through the climb, the plane would release the bomb, which would then travel upward, forward, and then downward onto the target. That delay of the detonation was sufficient for the bomber to complete its climb and turn, and to then move away from the explosion. This acrobatic maneuver is possible for modern high-performance jet fighter-bomber aircraft (atomic bombs with yields up to several hundred kilotons can be safely delivered in this fashion) but not for the massive, piston-powered B-29 aircraft that bombed Hiroshima and Nagasaki in 1945. Later in the book I'll discuss the mathematical physics of the alternative escape maneuver used by the slow, heavy bombers in the 1945 attacks. There are numerous YouTube videos on the Web of toss bombing, made with the aid of the Digital Combat Simulator (DCS) code available through Amazon for PCs.

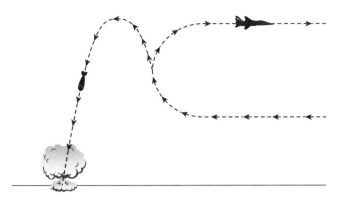

Figure P5.1. Toss-bombing and escape from an atomic explosion.

PROBLEM 6
Another Math Problem: This One Requires Calculus

Let's take a break from physics and do another purely math problem, one involving the sort of technical detail that mathematical physicists regularly encounter. If someone asks you which is larger, p^q or q^p, where p and q are both positive integers, you could easily answer that question by simply doing some multiplications. For example, which is greater, 4^3 or 3^4? Well, $4^3 = (4)(4)(4) = 64$, while $3^4 = (3)(3)(3)(3) = 81$, and so $3^4 > 4^3$. That was easy. But suppose one (or both) of p and q are *not* integers, or even rational; perhaps they are even *transcendental*. Things are not so straightforward now. Which is greater, e^π or π^e? You can't multiply e by itself π times, or multiply π by itself e times. And how about e^3 and 3^e, or 3^π and π^3, or e^2 and 2^e? NO CALCULATORS ALLOWED! There *is* an analytical way to quickly answer all four of these questions. As a hint to get you started, recall the infinite power series expansion of e^x, valid for all x (e is, of course, *Euler's number*: $\lim_{n \to \infty} \left(1 + \frac{1}{n}\right)^n = 2.71828\ldots$):

$$e^x = 1 + x + \frac{x^2}{2!} + \frac{x^3}{3!} + \ldots,$$

a formula that every mathematical physicist knows as well as his or her own birthday. Indeed, as a warm-up exercise, see if you can use the power series expansion to prove that $e < 3$. (It should be obvious from the series, *by inspection*, that $e > 2$; these two results show that e cannot be an integer.) If you can do that, and if you're feeling particularly frisky (in the mathematical sense), then see if you can further use the power series expansion to prove that e is irrational, that is, that there are no integers p and q such that $e = \frac{p}{q}$. The irrationality of e was established in 1737 by Euler.

PROBLEM 7
If Theory Fails: Monte Carlo Simulation

A mathematical physicist is occasionally confronted with a problem he or she can't solve analytically, either because there are no known techniques (yet) or simply because he/she *personally* doesn't know how to approach the problem mathematically. (There is no shame in this, as you can't know *everything*!) This doesn't necessarily mean all is lost, however, because there is a wonderful, additional tool available— the modern high-speed electronic computer.[1] Given the physics of a situation, a physicist may be able to get the answers he/she needs by what is called *Monte Carlo simulation* of that physics, instead of by solving the equations implied by the physics. Here's a simple example of what I'm getting at, in the form of a challenge problem first posed in the January 1878 issue of the American math journal *The Mathematical Visitor*.

Suppose a man walks across a square lawn with side length s. He does this by first picking, at random, a starting point on one edge

---

[1]Mathematicians, too, have learned to appreciate computers. As an example, consider a famous conjecture due to the great Euler, who, in 1769, asserted that there is no integer such that an nth power of it can be written as the sum of fewer than n nth powers of smaller integers. This conjecture defied formal analysis for two centuries until, in 1966, it was shown to be false by counterexample; that is, a computer search showed that, contrary to Euler, $144^5 = 27^5 + 84^5 + 110^5 + 133^5$, which has just *four* 5th powers on the right. Once in hand, this is "easily" confirmed by simply doing the arithmetic. I write "easily" somewhat loosely, as the arithmetic *is* a bit grubby, and so, because I know you're curious, the numbers are $61{,}917{,}364{,}224 = 14{,}348{,}907 + 4{,}182{,}119{,}424 + 16{,}105{,}100{,}000 + 41{,}615{,}795{,}893$.

of the lawn and then (again at random) picking an angle at which to walk with respect to that edge, with all angles from $0°$ to $180°$ being equally likely to be picked. What's the probability that the length of his path across the lawn is larger than s? This question *can* be answered theoretically, with the result being that the probability is $1 - \frac{2}{\pi} = 0.36338\ldots$, but what if we don't know enough probability theory to calculate this result?[2] What we'll do then is *simulate* the walk *10 million*(!) times and simply *count* how many of those walks have a length greater than the edge length. In Figure P7.1 I've shown the three distinct general paths our lawn-walker can take, on a square lawn with an edge length of unity. (Assuming $s = 1$ loses no generality, as all we care about is the *fraction* of paths with length greater than the edge length.) I've positioned the lawn in the first quadrant of a rectangular coordinate system, with the walker's starting point distance x from the origin, and the walk itself at angle θ to the bottom horizontal edge. Both x and θ are random, with x uniformly distributed from 0 to 1, and θ uniformly distributed from 0 to π radians. The path length is l (from 0 to $\sqrt{2}$), and we wish to find the probability that $l > 1$. From the figure we can see from elementary geometry that

(1) If the walk exits the right vertical edge, then

$$\frac{1-x}{l} = \cos(\theta), \text{ or } l = \frac{1-x}{\cos(\theta)};$$

(2) If the walk exits the top edge, then

$$\frac{1}{l} = \sin(\theta), \text{ or } l = \frac{1}{\sin(\theta)};$$

---

[2] No solution appeared in *The Mathematical Visitor*, but you can find a complete theoretical analysis in my book *Duelling Idiots and Other Probability Puzzlers*, Princeton University Press, 2002, pp. 58–61 and pp. 147–155. Most mathematical physicists *do* know enough probability theory to solve this particular problem theoretically, but there will *always* be *some* problem that defeats *anyone*, no matter how talented. That's when Monte Carlo can be the key to success.

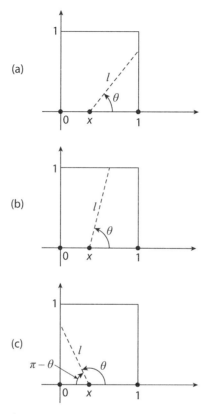

Figure P7.1. Random walks across the unit square.

(3) If the walk exits the left vertical edge, then

$$\frac{x}{l} = \cos(\pi - \theta) = -\cos(\theta), \ \text{or} \ l = -\frac{x}{\cos(\theta)}.$$

Case (a) occurs if

$$\theta < \tan^{-1}\left(\frac{1}{1-x}\right),$$

while case (c) occurs if

$$\theta > \pi - \tan^{-1}\left(\frac{1}{x}\right).$$

Case (b) occurs if

$$\tan^{-1}\left(\frac{1}{1-x}\right) < \theta < \pi - \tan^{-1}\left(\frac{1}{x}\right).$$

So, given the values of x and θ, we can use these conditions to first determine which case we have (that is, which edge the path exits) and then, from the appropriate l-equation, we can compute the path length. What we'll do, then, is perform this process a large number of times (10 million, in fact) and keep track of how many of those paths have $l > 1$. That's a *lot* of number crunching, but *we* aren't going to do it—our *computer* will do all the grubby arithmetic! The following computer code, **lawn.m**, does the job. It's written in MATLAB, but all the commands are, I believe, incorporated in all scientific programming languages, and so, I think—as I asserted in the preface—the code is virtually 100% transportable (or, at worst, requires only trivial modification). The one command that might require a comment is *rand*, which each time it is executed produces a number from a uniform distribution from 0 to 1. (The initial command ensures that the code will use a different sequence of random numbers every time the program is run.) When run multiple times, **lawn.m** estimated the probability we are after to be in the interval 0.36322 to 0.36348, a *narrow* interval that is centered on the theoretical value.

```
%lawn.m/created by PJNahin for Mathematical Physics
(5/27/2016)
rand('state',100*sum(clock)); paths=0;
for n=1:10000000
    x=rand;angle=pi*rand;
    A1=atan(1/(1-x));A2=pi-atan(1/x);
    if angle<A1
        l=(1-x)/cos(angle);
    elseif angle>A2
        l=-x/cos(angle);
    else
```

(continued)

```
                                          (continued)
        l=1/sin(angle);
    end
    if l > 1
        paths=paths+1;
    end
end
paths/10000000
```

Now that you've seen how Monte Carlo works,[3] here are three challenge problems for you to tackle. Of course, if you can do them theoretically, that's great, but even if you can do so, perform a Monte Carlo study, too, to check your theoretical work. Here's the first one: Imagine a circular archery target with three arrows stuck in it. If the arrows hit the target "at random,"[4] then what is the probability that no two of them are separated by more than the radius of the target?

The second problem is a bit more abstract. Two points are taken at random inside a quadrant of a circle, and a line is drawn through them. Find the probability (a) the line intersects the quarter-circle arc in two points, (b) the line intersects the quarter-circle arc in one point, and (c) the line does not intersect the quarter-circle arc. This problem received a theoretical study that I'll tell you about in the solution section. For both problems, your task is to discover an algorithm that describes the problem, and then use that algorithm to write and execute a computer code that estimates the probabilities to be found.

---

[3] For even more examples of Monte Carlo, see my book *Number-Crunching: Taming Unruly Computational Problems from Mathematical Physics to Science Fiction*, Princeton University Press, 2011.

[4] To generate points uniformly distributed over a circle, here's a particularly simple approach (physically clear, even if computationally a bit lengthy). Enclose the circle in a square with edge length equal to the circle's diameter, and then generate points uniformly distributed over the square (the points that happen to be inside the circle will then be uniform over the circle, as well). Use, of course, only those points that happen to fall inside the circle! For more on this, see my book *Digital Dice: Computational Solutions to Practical Probability Problems*, Princeton University Press, 2008, pp. 16–18.

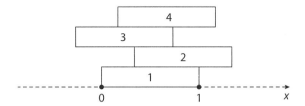

Figure P7.2. A random stacking of four bricks.

Those first two challenge problems are, admittedly, more mathematical than physical, and so, for a third challenge, here's a *physics* problem that I think you'll find presents a significant challenge to do theoretically but one (with a little thought) that is not at all difficult to simulate. Imagine that you have four identical rectangular bricks of uniform density, each of unit length. Imagine further that you *randomly*[5] stack the bricks (as shown in Figure P7.2), one brick at a time, as if you are building a wall. As you add each new brick the stack will either remain standing, or the new brick (and perhaps others) will fall. What is the probability that the stack is intact after the fourth brick has been added?

This problem was first posed by the American mathematician Artemas Martin (1835–1918) in the December 1871 issue of the *Schoolday Visitor*, with no solution given. It did, later, receive theoretical analysis, about which I'll tell you more in the solutions section. *Hint*: The key idea is, of course, based on the concept of *center of mass*. The underlying physics is elementary: (1) the center of mass of each individual brick is at the midpoint of the brick, and (2) the center of mass of the multiple masses $m_1, m_2, \ldots$, located at $x = x_1, x_2, \ldots$, respectively, is at

$$x = \frac{m_1 x_1 + m_2 x_2 + \ldots}{m_1 + m_2 + \ldots}.$$

[5]The bottom brick always has its left edge at $x = 0$ and its right edge at $x = 1$, as shown in Figure P7.2. A *random stacking* means that as you add each new brick to the stack, the right edge of the new brick could be directly over the left edge of the brick below it at one extreme, to the other extreme of having the left edge of the new brick directly over the right edge of the brick below it. A new brick's position is uniformly distributed over all possible positions between those two extremes.

For each brick in an intact stack, the center of mass of the bricks above it must be over that brick. If you want to try your hand at a simpler version of this question as a warm-up exercise, suppose you stack just three bricks and ask what is the probability the stack is intact after adding the third brick? A computer simulation should give you the result of about 0.2188.

PROBLEM 8

Monte Carlo and Theory: The Drunkard's One-Dimensional Random Walk

In the previous Monte Carlo problem a fair amount of computation was required to simulate the physics. There are, however, many fascinating physical situations that can be simulated with very little or almost no computation. Prime examples of this are what mathematicians call *random walks*. That term may sound vaguely frivolous, but in fact random walks occur in many important applications. Indeed, the origin of random walk theory (if not the term) can be traced as far back as 1904, to an analysis by the British medical doctor Sir Ronald Ross (1857–1932) of the random flights of mosquitoes as the airborne vectors of disease (yellow fever and malaria, in particular).[1]

The following year the great English statistician Karl Pearson (1857–1936) gave the picturesque name *drunkard's walk* to perhaps the most famous random walk of all (or, at least, to the random walk that students of probability theory in mathematics, and of statistical mechanics in physics, first encounter in their studies). The drunkard's walk, in one dimension, is as follows, illustrated in Figure P8.1.

Imagine a man who, after a night of wild celebration, wakes up in a drunken stupor at one of $N+1$ integer-valued points along the nonnegative x-axis ($x = 0, 1, 2, 3, \ldots, N$). If $1 \leq x < N - 1$, then once each unit of time thereafter he staggers one point to the right (with probability p) or one point to the left (with probability q). Since he

---

[1] You can find an extended discussion of the historical origin of the interest in random walks in my book *Mrs. Perkins's Electric Quilt and Other Intriguing Stories of Mathematical Physics*, Princeton University Press, 2009, pp. 233–284.

Figure P8.1. The drunkard's random walk.

will, after each time interval, stagger one way or the other, we know that $p + q = 1$. If he should ever reach $x = 0$ (the location of a pub), he stays there. If, however, he ever reaches $x = N$ (the location of his home), he goes to bed. The drunkard's walk problem is to calculate $f_N(x)$, the probability that he reaches home before he reaches the pub, given that he wakes up at location x.

From the problem statement we can immediately write

$$f_N(0) = 0, \quad f_N(N) = 1.$$

That's because if the drunkard wakes up in the pub ($x = 0$), then he stays there, and so there is zero probability he reaches home, and if he wakes up at home ($x = N$), then it is *certain* he reaches home. But what if he wakes up at some point other than the pub or home? In that case, we can write

$$f_N(x) = q f_N(x - 1) + p f_N(x + 1).$$

This equation (called a *second-order difference equation*) has a simple physical explanation. The probability of reaching home from position x (that is, $f_N(x)$) is the probability of reaching home from position $x - 1$ after moving (with probability q) to the left from position x to reach position $x - 1$, plus the probability of reaching home from position $x + 1$ after moving (with probability p) to the right from position x to reach position $x + 1$. So, there's the mathematical problem: solve

$$f_N(x) = q f_N(x - 1) + p f_N(x + 1), \ f_N(0) = 0, \ f_N(N) = 1$$

for $f_N(x)$. So, how do we do that?

It turns out that it is *easy* to do *if* you make the following observation: every term in the difference equation has $f_N(x)$, or a shifted version of $f_N(x)$, in it, and that strongly suggests a solution involving an exponential. That is, with s and C some constants, we might suspect[2] the solution will have the form $f_N(x) = Ce^{sx}$. See if you can take this hint and use it to solve the difference equation. (You should arrive at a *quadratic equation* for e^s, and so once again the authors of the *Boston Globe* letter I quoted in the preface might want to take notice.) Pay particular attention to the $p = q = \frac{1}{2}$ case (the so-called symmetrical walk).

But even if you can't solve the difference equation, you should now be able to appreciate how easy it is to simulate the drunkard's walk for any given values of p, $q (= 1 - p)$, and N. To check your simulation, in the solution section I'll show you how to solve the difference equation, which says (for example) that if $N = 11$ and $p = 0.6$, then $f_{11}(7) = 0.952$. Your assignment, then, is to write a Monte Carlo simulation for this walk, use it to confirm the theoretical value for $f_{11}(7)$, and then use the simulation to determine $f_{11}(x)$ for all the other values of x from 1 to 10.

PROBLEM 9
More Monte Carlo: A Two-Dimensional Random Walk in Paris

An obvious way to extend the one-dimensional random walk is to add a dimension to it, that is, to allow the random walker to be able to go not only left and right in the x-direction but also up and down in the y-direction. And, indeed, why stop there? One can continue to imagine adding a third, a fourth, and even more dimensions to the walk, with each additional dimension adding two more possibilities for the walker when choosing the next step. So, here's a two-dimensional drunkard's random walk for you to think about. I don't know a theoretical solution (if you develop one, please send it to me), but, just as with the one-dimensional drunkard's walk, it is easy to simulate on a computer:

---

[2] *Why* do we suspect this? Because you've seen problems like this before (and if you haven't, *now* you have!). The very first person to have seen this trick was *very* smart. All those who follow in the footsteps of that pioneer, however, don't have to be quite so smart.

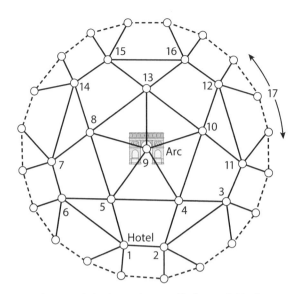

Figure P9.1. A random walk through Paris.

A drunken tourist starts at his hotel and walks at random through the streets of the idealized Paris shown in Figure P9.1. Find the probability that he reaches the Arc de Triomphe before he reaches the outskirts of town.[1]

In Figure P9.1 the nodes (the beginning and end points of a step) have been arbitrarily numbered from 1 (the hotel) to 16 (the Arc is node 9). All the outside nodes (the ones connected by the dashed circle) represent the outskirts of town and have the same number (17). So, a Monte Carlo simulation starts a large number of walks from node 1 and determines the fraction of those walks that reach node 9 before reaching node 17. As a partial check on the correctness of your code, explain why if you start a walk at node x (where $1 \leq x \leq 16$) and if $p(x)$ is the probability of reaching node 9 before reaching node 17, it

[1]Taken from the classic monograph by Peter G. Doyle and J. Laurie Snell, *Random Walks and Electric Networks*, The Mathematical Association of America, 1984, p. 59. This beautiful little book is an elegant illustration of the hand-in-hand relationship between mathematics (in particular, probability theory) and physics (in particular, electric circuit theory). For another example of using random walks to "solve" electrical circuit problems, see *Mrs. Perkins's Electric Quilt* (note 1 in Problem 8), pp. 299–320.

immediately follows that $p(1) = p(2)$, $p(4) = p(5)$, $p(3) = p(6)$, $p(7) = p(11)$, $p(8) = p(10)$, $p(12) = p(14)$, and $p(15) = p(16)$. Don't overlook the fact that, with two exceptions, each node has five exit paths, including those nodes that lead to the outskirts of town (each of those nodes has *two* such exit paths).[2] All the exit paths are equally likely; that is, each has probability $\frac{1}{5}$ of being chosen.

The central challenge to developing a Monte Carlo algorithm for this problem is devising a way to "tell the computer" how the nodes connect together. (Computer scientists call this the *data structure problem*, and it represents about 99% of the task. Once you've done that, you're practically done.) The fact that each node offers the same number of exit paths greatly helps simplify the data structure problem. An interesting extension of the data structure problem is how to handle multidimensional walks with varying numbers of exit paths for each node.

PROBLEM 10
Flying with (and against) the Wind:
Math for the Modern Traveler

Before getting to the main problem statement, here's a little warm-up for you. One day Bob runs an 8-mile course, with the first half of his run at 12 mph, and the second half at 8 mph. The next day Sally decides to run the same course; she also runs the first half at 12 mph and the second half at 8 mph. The time for Bob's run is *not* the same as the time for Sally's run. Explain how this can be, and calculate each runner's time.

Okay, now that you are all tuned up, here's a slightly more involved situation. An airline offers roundtrip travel between two cities, **A** and **B**, that are distance D apart. A plane that travels with a fixed ground speed of S in still air flies from **A** to **B** one day and then makes the return flight the next day. On the first leg of the roundtrip, the plane leaves **A** when the wind, blowing from **A** to **B**, is initially zero. At the

---

[2]The exceptions are, of course, node 9 (the Arc) and node 17, each of which terminates the walk.

start of the flight the wind speed steadily increases at a constant rate, so that when the plane arrives at **B** the wind speed is W. The next day, on the return flight from **B** back to **A**, the wind speed has fallen to a constant speed of $\frac{W}{2} < S$ but still blows from **A** to **B**. Compare the total roundtrip flight time of the airplane to what it would be in the absence of any wind. In particular, if $W = \frac{1}{2}S$, how different are the two total flight times (that is, which is the larger, and by how much)? This is an excellent example of the power of mathematical reasoning when applied to what appears to be a complicated physical situation.

Now, if you're feeling *really* powerful (in the mathematical sense), try your hand at the following.

Extra-Credit Challenge: A man runs along a straight path of length D at speed $v(t) = v(x)$. That is, his speed is $v(x)$ when he is distance x from his starting point $(x = 0)$, which he reaches at time t. Show that if T is the total time of the run, then

$$\frac{1}{T} \int\limits_0^T v(t)dt \leq \frac{1}{D} \int\limits_0^D v(x)dx.$$

That is, show that the *time*-average of his speed can never be greater than the *spatial*-average of his speed. *Hint*: Use the Cauchy-Schwarz inequality, which is discussed at the end of the preface. You will find it helpful, too, to remember that the relationship between D, T, and v is

$$D = \int\limits_0^T v(t)dt.$$

PROBLEM 11
**A Combinatorial Problem with Physics Implications:
Particles, Energy Levels, and Pauli Exclusion**

This discussion is intended to be an illustration of the interest mathematical physicists have in counting and in combinatorial analysis. Our particular interest here will be in what might appear to be, literally, child's play—the placing of balls in boxes—but which is actually an activity with a central role in atomic physics. In physics, the *balls* could be particles (electrons, neutrons, photons, etc.) and the *boxes* could be energy levels. To be very specific, suppose we have what mathematicians (and physicists, too) call *distinguishable* balls and *distinguishable* boxes. All the balls might, for example, be identical in every respect, but we can still tell them apart because each has a unique label pasted on it. If we have, again for example, three balls, let the labels be a, b, and c. Let's also suppose the boxes to be identical in every respect, but we can tell *them* apart as well because they have labels, too. If we have, for example, three boxes, let their labels be A, B, and C.

Our first question is, how many distinguishable ways are there to put the three balls into the three boxes, under the assumption that there is no limit on how many balls can go into a particular box? The answer is 27, and all those ways[1] are shown in Figure P11.1. What if the balls have their labels removed and so the balls are now *indistinguishable*? How many distinguishable ways are there now? The answer falls to 10, and all those ways are shown in Figure P11.2. What if, instead, the boxes are indistinguishable, while the balls are distinguishable? How many distinguishable ways are there to put the balls into the boxes? The answer is now even fewer, 5, and all of those ways are shown in Figure P11.3. Finally, what if both the balls *and* the boxes are indistinguishable? How many distinguishable ways to distribute the balls are there now? The answer is 3, and they are shown in Figure P11.4. The first two cases occur sufficiently often in physics that they have been given names (they are called *Maxwell-Boltzmann* and *Bose-Einstein* distributions, respectively),[2] names that mathematicians have also adopted.

---

[1]The reasoning is as follows. The first ball to be placed can go into any one of three boxes. The second ball to be placed can go into any one of three boxes. The third ball to be placed can go into any one of three boxes. So, the total number of distinguishable ways to distribute the balls is $(3)(3)(3) = 27$. In general, if we have n distinguishable boxes and r distinguishable balls, the total number of distinguishable ways to distribute the balls is n^r.

[2]After the Austrian physicist Ludwig Boltzmann (1844–1906), the Scottish physicist J. C. Maxwell (1831–1879), the Indian physicist S. N. Bose (1894–1974), and (of course!) the German-born physicist Albert Einstein (1879–1955).

1. $\{abc|\text{—}|\text{—}\}$ **10.** $\{a|bc|\text{—}\}$ **19.** $\{\text{—}|a|bc\}$

2. $\{\text{—}|abc|\text{—}\}$ **11.** $\{b|ac|\text{—}\}$ **20.** $\{\text{—}|b|ac\}$

3. $\{\text{—}|\text{—}|abc\}$ **12.** $\{c|ab|\text{—}\}$ **21.** $\{\text{—}|c|ab\}$

4. $\{ab|c|\text{—}\}$ **13.** $\{a|\text{—}|bc\}$ **22.** $\{a|b|c\}$

5. $\{ac|b|\text{—}\}$ **14.** $\{b|\text{—}|ac\}$ **23.** $\{a|c|b\}$

6. $\{bc|a|\text{—}\}$ **15.** $\{c|\text{—}|ab\}$ **24.** $\{b|a|c\}$

7. $\{ab|\text{—}|c\}$ **16.** $\{\text{—}|ab|c\}$ **25.** $\{b|c|a\}$

8. $\{ac|\text{—}|b\}$ **17.** $\{\text{—}|ac|b\}$ **26.** $\{c|a|b\}$

9. $\{bc|\text{—}|a\}$ **18.** $\{\text{—}|bc|a\}$ **27.** $\{c|b|a\}$

Figure P11.1. The 27 distinguishable ways to put 3 distinguishable balls into 3 distinguishable boxes.

1. $\{***|\text{—}|\text{—}\}$ **6.** $\{*|**|\text{—}\}$

2. $\{\text{—}|***|\text{—}\}$ **7.** $\{*|\text{—}|**\}$

3. $\{\text{—}|\text{—}|***\}$ **8.** $\{\text{—}|**|*\}$

4. $\{**|*|\text{—}\}$ **9.** $\{\text{—}|*|**\}$

5. $\{**|\text{—}|*\}$ **10.** $\{*|*|*\}$

Figure P11.2. The 10 distinguishable ways to put 3 indistinguishable balls into 3 distinguishable boxes.

1. $\{a|b|c\}$ **2.** $\{abc|\text{—}|\text{—}\}$ **3.** $\{ab|c|\}$ **4.** $\{bc|a|\text{—}\}$ **5.** $\{ac|b|\text{—}\}$

Figure P11.3. The 5 distinguishable ways to put 3 distinguishable balls into 3 indistinguishable boxes.

1. $\{***|\text{—}|\text{—}\}$ **2.** $\{**|*|\text{—}\}$ **3.** $\{*|*|*\}$

Figure P11.4. The 3 distinguishable ways to put 3 indistinguishable balls into 3 indistinguishable boxes.

In some physics applications a restriction, called the *Pauli exclusion principle*,[3] is applied, which says that no box can contain more than one ball. So, how many distinguishable ways are there to put three balls into

---

[3]After the Austrian-born Swiss and American physicist Wolfgang Pauli (1900–1958).

1. $\{a\,|\,b\,|\,c\}$ **2.** $\{a\,|\,c\,|\,b\}$ **3.** $\{b\,|\,a\,|\,c\}$

4. $\{b\,|\,c\,|\,a\}$ **5.** $\{c\,|\,a\,|\,b\}$ **6.** $\{c\,|\,b\,|\,a\}$

Figure P11.5a. The 6 distinguishable ways to put 3 distinguishable balls into 3 distinguishable boxes *with Pauli exclusion*.

1. $\{*\,|\,*\,|\,*\}$

Figure P11.5b. The lone distinguishable way to put 3 indistinguishable balls into 3 distinguishable boxes *with Pauli exclusion*.

three distinguishable boxes under the Pauli exclusion constraint? If the balls are distinguishable, the answer is 6, as shown in Figure P11.5a (the last entries in Figure P11.1), while if the balls are indistinguishable the answer is 1, as shown in Figure P11.5b (the last entry in Figure P11.2). This last case is called a *Fermi-Dirac* distribution.[4]

To specifically list all the various ways of placing balls in boxes under the various restrictions (or not) of distinguishability, indistinguishability, and Pauli exclusion *very* quickly becomes hugely tedious as the number of balls and/or boxes increases beyond three. (Try four balls in five boxes!) What we need are general formulas that will give us the number of ways to do it, given whatever restrictions apply. As an example of doing that, in how many ways can you place n distinguishable balls in n distinguishable boxes, assuming the Pauli exclusion principle does *not* apply? We can argue as follows, taking each ball one at a time. The first ball can go into any one of the n boxes. The second ball can go into any one of the n boxes, and so on, for all the rest of the balls. So, there are n^n distinguishable ways to put n distinguishable balls into n distinguishable boxes if Pauli exclusion does not apply (see footnote 1 again, with $r = n$).

An interesting follow-up question now presents itself: how many of those n^n ways have exactly one ball in each box? The first box can receive any one of the n balls. The second box can receive any one of the remaining $n-1$ balls. The third box can receive any one of the remaining $n-2$ balls, and so on, until we get to the

---

[4]After the Italian-born American physicist Enrico Fermi (1901–1954) and the English physicist Paul Dirac (1902–1984).

last box, which receives the now lone remaining ball. So, there are $n(n-1)(n-2)\ldots(1) = n!$ distinguishable ways to put exactly one ball in each of the boxes. This tells us that the probability of there being exactly one ball in each box is

$$\frac{n!}{n^n},$$

a result with some nonintuitive implications.

For example, suppose we throw a fair die—with (of course) six equally likely distinguishable faces—six times. We consider each toss as a "box," and each face that shows on a toss as a "ball." So, the probability that in six tosses each face of the die shows exactly once is

$$\frac{6!}{6^6} = 0.0154\ldots,$$

which is surprisingly small. If we call the tossing of the die six times an "experiment," then, *on average*, we'd expect to see all six faces only once every 65 experiments. All the other experiments give one (or more) of the faces occurring multiple times. That is, in a truly random experiment (like the tossing of a fair die six times), *clustering* or *lumping* is the expected outcome, not the uniform distribution most people think "intuitively obvious."

What's the probability that one of the faces shows multiple times, and different faces show on the remaining tosses (in a total of six tosses)? This says one of the faces (let's say the ace) doesn't show at all, and so for each toss there are only five possibilities, for a total of 5^6 possibilities in which no ace shows. Since there are 6^6 possibilities in general, the probability that no ace shows in six tosses (which means some other face shows at least twice) is

$$\frac{5^6}{6^6} = 0.33489\ldots,$$

which is nearly 22 times as likely as is the uniform case of each face showing once in six tosses.

As another example of the problem of putting n distinguishable balls into n distinguishable boxes with no box empty, suppose $n = 365$.

This has the amusing interpretation of asking what is the probability, given 365 randomly selected people, that every day of the year is a birthday for somebody? Intuitively, you probably think (correctly) that the answer is pretty small—but *how* small? To answer that, all we need do is calculate the value of

$$\frac{365!}{365^{365}},$$

—and lots of luck doing that on your run-of-the-mill hand calculator! What we'll use instead is *Stirling's asymptotic expression for the factorial*,[5] which is

$$n! \sim \sqrt{2\pi n}\, n^n e^{-n}.$$

So,

$$\frac{365!}{365^{365}} \sim \frac{\sqrt{2\pi 365}\, 365^{365} e^{-365}}{365^{365}} \approx \frac{17{,}500}{(365e)^{365}} \approx \frac{1.75 \times 10^4}{992^{365}}$$

$$\approx \frac{1.75 \times 10^4}{(10^{2.9965})^{365}} \approx 1.75 \times 10^4 \times 10^{-1{,}094}$$

$$\approx 2 \times 10^{-1{,}090},$$

which, even if we are off by an order of magnitude or two or three, is (very!) small. (For more on this particular problem, see Problem 15.)

Finally, suppose we have r distinguishable balls that are placed in n distinguishable boxes. What is the probability that no box has more than one ball in it (that is, every box is either empty or contains at

---

[5] Named after the Scottish mathematician James Stirling (1692–1770)— although it is known that the French mathematician Abraham de Moivre (1667–1754) knew an equivalent form at the same time (or even earlier)—who published it in 1730. Factorials get very large, very fast (my hand calculator first fails at 70!), and Stirling's formula is quite useful in computing $n!$ for large n. It is called *asymptotic* because while the *absolute* error in the right-hand side in evaluating the left-hand side blows up as $n \to \infty$, the *relative* error goes to zero as $n \to \infty$ (that's why $\sim$ is used instead of $=$). That is, $\lim\limits_{n \to \infty} \frac{n!}{\sqrt{2\pi n}\, n^n e^{-n}} = 1$.

most just one ball)? If $r > n$, the answer is, of course, zero. For $r \leq n$ we reason as follows. The first ball can go into any one of the n boxes. The second ball can go into any one of the still-empty $n - 1$ boxes. The third ball can go into any one of the still-empty $n - 2$ boxes, and so on, down to the final rth ball, which can go into any one of the still-empty $n - r + 1$ boxes. So, the probability we are after is

$$\frac{n(n-1)(n-2)\ldots(n-r+1)}{n^r}.$$

This is, of course, a version of the famous so-called birthday problem, which asks for the probability that r randomly selected people will all have different birthdays. If we evaluate the preceding probability expression for $r = 22$ and $r = 23$, we find that the probability that all r people have different birthdays is

$$r = 22: \quad \frac{(365)(364)\ldots(344)}{365^{22}} = 0.5243,$$

and

$$r = 23: \quad \frac{(365)(364)\ldots(343)}{365^{23}} = 0.4927.$$

And so, for just $r = 23$ people, the probability is greater than $\frac{1}{2}$ that at least two people *will* have the same birthday. Most people find this rather surprising (surprising, that is, that the required r is so small). When I taught introductory probability theory to electrical engineering undergraduates at the University of New Hampshire, I made it a regular practice to start my first lecture with an experimental version of this problem. That is, after posing it to the class (generally with 35 to 40 students) and getting guesses typically ranging from 100 to 200 to achieve a probability of $\frac{1}{2}$ in getting a match, I'd then ask the students to call out their birthday (month and day), one after the other. Almost always there was a match, to the great delight of all (most of all, *me!*).

Okay, with all that said, here are three questions for you to think about. The first one is easy, and you should be able to answer it without hesitation: what is the probability of putting n distinguishable balls into n distinguishable boxes such that exactly one box remains empty, *if Pauli exclusion is in effect*? The second question is just a bit more difficult: how does the answer change if Pauli exclusion is *not* in effect? *Hint*: If you evaluate your answer for $n = 4$, you should get 0.5625, and for $n = 7$, you should get 0.128517.... And finally, for a third task, write a Monte Carlo simulation for the $n = 7$ case and see if computer "experiment" agrees with theory.

PROBLEM 12

Mathematical Analysis: By *Physical* Reasoning

If n is any nonnegative *integer*, it had been known for centuries before the time of the great English mathematical physicist Isaac Newton (1642–1727) that

$$(x+y)^n = \sum_{j=0}^{n} \binom{n}{j} x^{n-j} y^j = \binom{n}{0} x^n + \binom{n}{1} x^{n-1} y$$

$$+ \ldots + \binom{n}{n-1} xy^{n-1} + \binom{n}{n} y^n,$$

where

$$\binom{n}{j} = \frac{n!}{j!(n-j)!}$$

denotes the so-called binomial coefficients.[1] Indeed, this expansion of $(x+y)^n$ is called the *binomial theorem*, and it is commonly encountered in second-year high school algebra. At some point in 1664, while still an undergraduate at Cambridge, Newton came to understand that the theorem could be generalized to cover the case of n being *any* rational number, either positive or negative, and n is not confined to being a nonnegative integer.

---

[1]When working with factorials, remember that $0! = 1$ (*not* 0). That's because $j! = j(j-1)!$ and so, for $j = 1$, we have $1! = 1(0!)$ and, since $1! = 1$, then $0! = 1$. To be emphatic about this, $0! \neq 0$ (!!!).

Newton neither published nor even proved the generalized theorem but simply used it. It wasn't, in fact, until the early part of the 19th century that it was finally proven by the Norwegian mathematician Niels Henrik Abel (1802–1829), in 1826.[2] Our interest here, however, isn't in the proof of the theorem but rather in the binomial coefficients $\binom{n}{j}$ themselves. The coefficients occur in physics in numerous (and often surprising) applications, and mathematical physicists who are good at manipulating these *mathematical* entities often find that skill crucial in solving their *physics* problems.[3] Indeed, understanding the *physics* behind what $\binom{n}{j}$ *means* is fundamental.

To start, notice that I simply asserted earlier that for n and j positive integers,

$$\binom{n}{j} = \frac{n!}{j!(n-j)!}.$$

But what does $\binom{n}{j}$ *physically* mean? Imagine you have a pile of n blocks in front of you, and you wish to select j of them to put in a sack. The claim is that the number of ways you can do that *is* $\binom{n}{j}$. Notice that *if* this is so (I'll show you *why* it is so in just a moment), then we immediately have, *physically*, that for all $n \geq 0$

$$\binom{n}{j} = 0$$

for both $j > n$ and $j < 0$. That's because it's impossible to put more blocks in the sack than there are blocks to start with, and it's also impossible to put fewer than no blocks in the sack. (That's *physics*, not mathematics.)

---

[2]While he didn't formally publish the generalized binomial theorem, neither did Newton keep it to himself as a secret. For example, he told the German mathematician Gottfried Leibniz (1646–1716) all about it in an exchange of letters. See James R. Newman, *The World of Mathematics* (vol. 1), Simon and Schuster, 1956, pp. 519–524.

[3]A good illustration of this can be found in the work of the American mathematician John Riordan (1903–1988), who worked for decades at the Bell Telephone Laboratories. Riordan's classic 1968 book *Combinatorial Identities* (Wiley) shows the sort of mathematics he encountered in his analyses of electronic switching and information routing networks (telephone circuits).

Now, *mathematically*, first there are n choices for the first block to go in the sack. Then there are $n-1$ choices for the second block to go in the sack, and so on, down to $n-j+1$ choices for the jth (and last) block to go in the sack. So, the total number of ways (where the *order* of selection matters) to select j blocks from n blocks is

$$n(n-1)(n-2)\ldots(n-j+1)$$
$$= \frac{[n(n-1)(n-2)\ldots(n-j+1)][(n-j)(n-j-1)\ldots(3)(2)(1)]}{[(n-j)(n-j-1)\ldots(3)(2)(1)]}$$
$$= \frac{n!}{(n-j)!}.$$

But if the order of selection *doesn't* matter—if we don't care about the order in which those particular j blocks go into the sack but, rather, only that those particular j blocks are the ones in the sack at the end of the selection process—then we need to divide this last expression by the number of ways those particular j blocks could have been selected. That is, by $j!$, because there are j choices for the first block selected, $j-1$ choices for the second block selected, and so on.

So, the number of ways to select j blocks to go into a sack, from n blocks, *where the order of selection doesn't matter,* is

$$\frac{n!}{j!(n-j)!},$$

which we write, in shorthand, as $\binom{n}{j}$.

The binomial coefficients satisfy numerous identities, expressions with which mathematical physicists need to be comfortable. One simple identity that follows instantly from the very mathematical definition of $\binom{n}{j}$ is that

$$\boxed{\binom{n}{j} = \binom{n}{n-j}.}$$

Physically, this is the statement that the number of ways to select j blocks to put in a sack from n blocks is equivalent to selecting the $n-j$ blocks *not* to put in the sack.

Here's another identity that is equally easy to establish:

$$\sum_{j=0}^{n} \binom{n}{j} = 2^n.$$

To see this, simply set $x = y = 1$ in the binomial theorem and then notice that 1 raised to any power is still 1. And here's yet another identity, probably one of the most useful of all:

$$\binom{n+1}{j} = \binom{n}{j} + \binom{n}{j-1}.$$

You can prove this by pure algebraic manipulation (it's not particularly difficult—try it!). But what I'll show you now is the *physicist's* way of establishing this identity, which gives it a reality above and beyond being simply a bunch of abstract mathematical symbols.

Suppose you have $n + 1$ books in your home library, and you want to select j of them to put in your travel bag as you leave for vacation. From what has been said before, the number of ways to do that is $\binom{n+1}{j}$. We can also calculate the number of ways using the following alternative, two-step argument. Suppose that before you start your selection process, you tie a pink ribbon around one of the $n + 1$ books, a book you choose at random. Then, when you are done selecting the j books, exactly one of two things will have occurred: (1) the book with the pink ribbon will *not* be in the bag, or (2) the book with the pink ribbon *will* be in the bag. If it's (1), then you picked *all* the j books from the n books without the ribbon, which can happen in $\binom{n}{j}$ ways. If it's (2), then there are $j - 1$ books in the bag (along with the one with the pink ribbon) that were selected from n books, which can happen in $\binom{n}{j-1}$ ways. Since (1) and (2) cover all the different ways that could possibly happen, we have our identity.

Here's another "physics" derivation of a famous mathematical result. Suppose you have n black balls and m white balls, all in a box. How many ways can you select j of those balls to put in a bag? The balls are distinguishable (all are uniquely numbered, for example). Thus, we immediately know that the answer is $\binom{n+m}{j}$. But we can also calculate the answer in a different way, as follows. Suppose k of the j balls are black, which means that $j - k$ balls are white. We can select the k black balls in $\binom{n}{k}$ ways, and the $j - k$ white balls in $\binom{m}{j-k}$ ways. So, there are $\binom{n}{k}\binom{m}{j-k}$ ways to select k black balls and $j - k$ white balls. Since k can vary from 0 to j, we have a total of

$$\sum_{k=0}^{j} \binom{n}{k}\binom{m}{j-k}$$

ways, which of course must equal $\binom{n+m}{j}$. That is,

$$\binom{n+m}{j} = \sum_{k=0}^{j} \binom{n}{k}\binom{m}{j-k},$$

which is called the *Vandermonde identity*.[4]

The Vandermonde identity can be used to establish other identities that appear, at least at first glance, to be quite exotic. Consider, for example,

$$\sum_{j=0}^{k} \binom{k}{j}^{2} \binom{n + 2k - j}{2k} = \binom{n + k}{k}^{2},$$

which can be established by a *triple* application of the Vandermonde identity.[5] This identity has been traced back to an 1867 book, often attributed incorrectly to a Chinese mathematician named "Le-Jen Shoo," itself a botched version of "Le-Jen Shu." But that's not right either: the correct attribution is to Li Shanlan (1810–1882), a well-known mathematician of the Qing dynasty.

[4]After the French mathematician Alexandre-Théophile Vandermonde (1735–1796), who published it in 1772. It was, however, known to Chinese mathematicians long before that.

[5]For details, see a brief note by T. S. Nanjundiah in the *American Mathematical Monthly*, May 1958, p. 354.

Setting $m = j = n$ in the Vandermonde identity gives us another particularly interesting special case:

$$\binom{2n}{n} = \sum_{k=0}^{n} \binom{n}{k} \binom{n}{n-k} = \sum_{k=0}^{n} \binom{n}{k} \binom{n}{k},$$

because, you'll recall,

$$\binom{n}{n-k} = \binom{n}{k}.$$

That is,

$$\binom{2n}{n} = \sum_{k=0}^{n} \binom{n}{k}^2.$$

I call this, and the Li Shanlan identity, interesting and exotic because the *squares* of binomial coefficients appear. Identities involving powers of binomial coefficients have a touch of mystery to them, and so, when a new one is discovered, it quickly becomes part of an elite collection. Here's another famous example of that, one for which I don't know a "physics derivation":

$$\sum_{k=0}^{2n} (-1)^k \binom{2n}{k}^3 = (-1)^n \frac{(3n)!}{n!^3}.$$

This beautiful expression, called *Dixon's identity*,[6] involves *cubes* of binomial coefficients, and if you discover a "physics derivation" for it, please write and tell me all about it!

Okay, here are *your* problems to ponder. First, you'll recall that I invoked physics in support of declaring $\binom{n}{j} = 0$ for both $j > n$ and

---

[6]After the Irish mathematician Alfred Cardew Dixon (1865–1936), who published it in 1891. Dixon's original proof, and all those that have since appeared, are highly mathematical in nature. A "physics derivation" would be of great interest, even to mathematicians.

$j < 0$. Thus, you might think that, even for $0 \le j \le n$, $\binom{n}{j} = 0$, too, if $n < 0$. That is, a physicist would surely argue (*you'd think*) that it's simply impossible to select j blocks from *fewer than none*! But that's not an argument that stops a pure mathematician. Writing, for x *any* real number,

$$\binom{x}{r} = \frac{x(x-1)(x-2)\ldots(x-r+1)}{r!},$$

where r is a positive integer,[7] it then quickly follows that

$$\binom{-1}{r} = (-1)^r, \quad \text{and} \quad \binom{-2}{r} = (-1)^r(r+1).$$

Explain these two claims, and then calculate $\binom{-3}{r}$. *Hint:* As a partial check on your answer, $\binom{-3}{7} = -36$, and $\binom{-3}{8} = 45$.

For a second challenge, find an expression for $\frac{\binom{1/2}{n}}{\binom{-1/2}{n}}$. *Hint:* For $n = 3$, your expression should give $-\frac{1}{5}$.

As a third challenge, consider the *Catalan numbers*, defined to be

$$C_n = \frac{1}{n+1}\binom{2n}{n} = \frac{(2n)!}{n!(n+1)!}, n = 0, 1, 2, 3, \ldots,$$

named after the Belgian mathematician Eugène Catalan (1814–1894), who encountered them during an 1838 analysis of a combinatorial problem. (Euler had, in fact, discovered them decades earlier in a different problem.) The C_n are integers (1, 1, 2, 5, 14, 132, 429, 1430, ...), although that may not be immediately obvious from the definition. See if you can develop a convincing demonstration that shows this. *Hint: Assume* binomial coefficients are integers,[8] and then show $C_n = \binom{2n}{n} - \binom{2n}{n+1}$.

---

[7]This expression for $\binom{x}{r}$, x any real number, reduces to our earlier $\frac{x!}{r!(x-r)!}$ if x is a positive integer.

[8]Mathematical physicists readily accept that binomial coefficients are integers because of their *physical* interpretation, but mathematicians want a purely *mathematical* proof that $\binom{n}{k} = \frac{n!}{k!(n-k)!}$ is an integer. This is not a difficult problem, but it is also not a trivial problem. If you try your hand at it, you may find Ball's result (see note 4 in the preface) helpful.

PROBLEM 13

When an Integral Blows Up: Can a Physical Quantity Really Be *Infinite?*

What does it mean if when you are doing a theoretical analysis, a *mathematical* analysis produces a result that is obviously *physically* wrong? One answer, of course, is simply that somewhere in the analysis you screwed up. All of us *can*, after all, now and then add 2 to 3 and get 4. But let's suppose you haven't done that. Instead, in fact, what if every step leading up to the clearly wrong physical result has been certified to be mathematically correct? What are we to make of *that*? This isn't a hypothetical question, as the history of physics is full of such surprises. Sometimes the way out has been a revolution in thought (quantum mechanics is a prime example), but often a resolution is a bit less dramatic.

As an example of what I'm talking about, in the next problem we will analyze an easy-to-understand physical process, and after some quite elementary geometry, we'll arrive at the integral

$$I(\alpha) = \int\limits_{0}^{\alpha} \frac{d\theta}{\sqrt{1 - \cos(\theta)}},$$

where the upper limit $\alpha > 0$. This quite simple-appearing integral will have a very specific physical interpretation, but that will present us with a real puzzle, because the integral *doesn't exist*; that is, $I(\alpha) = \infty$. The problem with this is that there is *nothing* in the physical world that is truly infinite, with the possible exception of whatever might be at the center of a black hole. Getting an infinity in a mathematical physics analysis is the very signature of something gone really bad. That *something* is what we'll be taking up in the next problem, but for now, the focus is on *how* we know $I(\alpha) = \infty$.

Now, it might be tempting to argue that the answer is that the integrand blows up at the lower limit; that is, $I(\alpha) = \infty$ because

$$\lim_{\theta \to 0} \frac{1}{\sqrt{1 - \cos(\theta)}} = \infty.$$

That limit *does* blow up, yes, but it's not a good argument for why $I(\alpha) = \infty$. That's because it's easy to produce a counterexample. For example, the integrand in

$$\int\limits_0^1 \frac{d\theta}{\sqrt{\theta}}$$

also blows up at the lower limit, but the integral itself is

$$\left\{2\sqrt{\theta}\right\}\big|_0^1 = 2,$$

which I'm pretty sure we'd all agree is finite. So, the answer to *why* $I(\alpha) = \infty$ requires a somewhat deeper explanation.

Well, okay, you say, let's quit fooling around and just *do* the blasted integral. *That'll* tell us if $I(\alpha) = \infty$. Yes, that would be definitive, all right, but to do that requires some work. (You, in fact, will eventually be asked to do that work—guided by some helpful hints from me!— as a challenge.) For right now, though, let me show you an easy way to understand *why* $I(\alpha) = \infty$, using a general approach that every mathematical physicist ought to have in his or her arsenal of tricks.

We start by observing that for θ any real-valued angle, $\cos(\theta) \leq 1$. Then, integrating this inequality from 0 to an arbitrary positive upper limit (call it x), we have

$$\int\limits_0^x \cos(\theta)d\theta \leq \int\limits_0^x d\theta,$$

which quickly leads to $\sin(x) \leq x$. Then, we integrate this inequality from 0 to an arbitrary upper limit (call it y) to get

$$\int\limits_0^y \sin(x)dx \leq \int\limits_0^y x\, dx,$$

which quickly leads to $\cos(y) \geq 1 - \frac{1}{2}y^2$.

Now, if in the $I(\alpha)$ integral we replace $\cos(\theta)$ in the denominator of the integrand with something *smaller*, that is, with $1 - \frac{1}{2}\theta^2$, we'll get a *larger* denominator and thus a *smaller* integrand. This will result in an integral *smaller* than $I(\alpha)$. So,

$$I(\alpha) = \int_0^\alpha \frac{d\theta}{\sqrt{1-\cos(\theta)}} \geq \int_0^\alpha \frac{d\theta}{\sqrt{\frac{1}{2}\theta^2}} = \sqrt{2}\int_0^\alpha \frac{d\theta}{\theta} = \sqrt{2}\{\ln(\theta)\}|_0^\alpha = \infty,$$

because the logarithm blows up to minus infinity at the lower limit of $\theta = 0$. So, $I(\alpha)$ is "larger than infinity," which is just an enthusiastic way of saying it, too, diverges. (We'll use this result in the next problem.)

Okay, now for *your* assignment: confirm this conclusion about $I(\alpha)$ by actually doing the integral, *exactly*. That is, see if the following sequence of hints allows you to conclude that $I(\alpha)$ does, indeed, blow up logarithmically.

(a) Make the change of variable $z = \tan\left(\frac{\theta}{2}\right)$, and remember the identity
$\tan^2\left(\frac{\theta}{2}\right) = \frac{1-\cos(\theta)}{1+\cos(\theta)}$;[1]

(b) Show that $\cos(\theta) = \frac{1-z^2}{1+z^2}$;

(c) Show that $\sin(\theta)\frac{d\theta}{dz} = \frac{4z}{\left(1+z^2\right)^2}$;

(d) Show that $\sin(\theta) = \frac{2z}{1+z^2}$;

(e) Show that $\theta = \frac{2}{1+z^2}dz$;

(f) Show that $I(\alpha) = \sqrt{2}\int_0^{\tan\left(\frac{\alpha}{2}\right)} \frac{dz}{z\sqrt{1+z^2}}$;

(g) Look up the last integral in a math table.

As a second challenge, in the text I derived the upper-bound inequality $\sin(x) \leq x$. See if you can discover a proof for this lower-bound inequality for $\sin(x)$: $\frac{2}{\pi}x \leq \sin(x)$, $0 \leq x \leq \frac{\pi}{2}$. *Hint*: Draw a picture and recall the equation for a straight line.

---

[1]Am I serious about remembering this identity? Well, no, but this is a good illustration of the value of having a well-thumbed math handbook in your personal library. You don't have to do *any* of this, of course, *if* you have a book of integrals that has $\int \frac{d\theta}{\sqrt{1-\cos(\theta)}}$ in it. All I have handy in my home library is the brief Schaum's integral tables, however—and it's the weekend as I type this and the university's math library is closed—and so this is what I did. Perhaps I should buy a bigger book of integrals!

PROBLEM 14
Is This Easier Than Falling Off a Log? Well, Maybe Not

In this discussion I'll show you an example of how computers can help mathematical physicists "understand" their equations, and not just work as a tool to *simulate* physics. The problem may at first appear to be simply an academic whimsy, but some quite serious physics is involved. The problem *is*, I must admit, a favorite of textbook authors precisely *because* of the whimsy; I suspect, in fact, that it has appeared in one form or another in just about every undergraduate physics text for *decades* (I remember doing a version of it, as a homework problem, in a theoretical mechanics class at Stanford nearly 60 years ago!). And yet, there is a subtle "flaw" to it that I have not seen discussed in the literature. It appears here, I believe (with a hint of it in the previous problem), for the first time.

Imagine a point mass m placed at the very top of a hemispherical blob of ice. The ice has been sprayed with water, and so, with a thin film of liquid water on it, the blob is *very* slippery. Indeed, let's assume the ice is without any friction at all. The textbook question is to calculate the angle θ (see Figure P14.1) at which the point mass (*with zero initial velocity*) will fly off the ice surface (along a path tangent to the hemispherical surface) as the mass slides downward. The mass *will* fly off *if it is sliding* (more on this crucial point, soon), because as it slides downward with ever-increasing speed, the inward centripetal acceleration force required to keep the mass on the surface will eventually exceed the available inward-directed radial component of the mass's weight. At that instant, the hemisphere's reaction force F_n to the mass's weight (mg)—where g is the acceleration of gravity—which is outwardly normal (perpendicular) to the hemisphere's surface, will have fallen to zero, and so the mass will then leave the surface of the ice.

Since the mass has zero radial motion until it reaches point P (where it leaves the hemisphere's surface), the inward radial force of gravity must provide, exactly, the sum of the magnitudes of the inward-directed centripetal force and the outward-directed reaction force. We

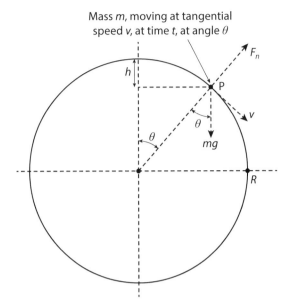

Figure P14.1. Sliding on a slippery hemisphere of ice.

can set all that up mathematically by writing

$$mg\cos(\theta) = \frac{mv^2}{R} + F_n,$$

where v is the tangential speed of the mass, and R is the radius of the hemisphere.

Another equation we can write is

$$m\frac{dv}{dt} = mg\sin(\theta),$$

which is Newton's famous second law of motion: the *tangential* acceleration force on the left-hand side is derived on the right-hand side by the *tangential* component of the force of gravity on the mass. Now, if we denote the angle θ at point P by θ_P (reached at time $t = t_P$, where $t = 0$ is the start of the slide), and the speed at point P by v_P, then set

$F_n = 0$, we have

$$mg \cos(\theta_P) = \frac{m v_P^2}{R},$$

and so

$$\cos(\theta_P) = \frac{v_P^2}{gR}.$$

By conservation of energy, when the mass has dropped through the vertical distance h (see Figure P14.1) it will have lost gravitational potential energy in the amount of mgh, *all* of which must exactly be found as an *increase* in the kinetic energy of the mass (with zero friction, there is no heat-generating energy-loss mechanism in the problem). By simple geometry we have, at point P,

$$h = R - R \cos(\theta_P) = R[1 - \cos(\theta_P)].$$

And so, as the initial kinetic energy is zero, we have

$$\frac{1}{2} m v_P^2 = mg R[1 - \cos(\theta_P)],$$

or

$$\frac{v_P^2}{gR} = 2[1 - \cos(\theta_P)] = 2 - 2 \cos(\theta_P),$$

and so, by the preceding boxed equation,

$$\cos(\theta_P) = 2 - 2 \cos(\theta_P).$$

That is,

$$\cos(\theta_P) = \frac{2}{3}.$$

(Notice, carefully, that if we drop the subscript P from v_P and θ_P, the last boxed equation continues to hold in general for v and θ at all points *prior* to reaching point P in Figure P14.1.)

Thus, just like that, we have the answer to the usual textbook question about the problem (what is the angle at which m flies off the surface of the hemisphere?):

$$\theta_P = \cos^{-1}\left(\frac{2}{3}\right) = 48.2°.$$

This *is* a remarkable result, as it is *independent* of m, R, and g. That is, big mass or little mass, big ice blob or little ice blob, low gravity or high gravity—none of any of that matters: the angle of departure remains the same in all cases. This feature, alone, accounts for the popularity of the problem for textbook and lecture use.

But there is another calculation we can do, one that I haven't seen in textbooks and one that raises a very puzzling question. That's the calculation of the *duration* of the slide until the mass is launched into space. If we write s as the distance traveled on the surface of the hemisphere by the point mass when it has traveled through angle θ, then we have

$$s = R\theta,$$

and so the tangential speed of the mass is

$$v = \frac{ds}{dt} = R\frac{d\theta}{dt};$$

that is,

$$dt = R\frac{d\theta}{v}.$$

As we showed earlier (see the preceding boxed equation), the speed of the sliding mass at angle θ is

$$v = \sqrt{2gR}\sqrt{1 - \cos(\theta)},$$

and so, at the instant $t = t_P$ (when the mass flies off the hemisphere) we have $\theta = \theta_P$, and thus

$$v(\theta_P) = v_P = \sqrt{2gR}\sqrt{1 - \frac{2}{3}} = \sqrt{\frac{2gR}{3}},$$

a well-defined expression for the fly-off speed. For example, if $R = 3$ meters and if $g = 9.81 \text{ m/s}^2$, then $v_P = 4.43 \text{ m/s}$.

But things don't go nearly so well for the fly-off *time*. Since

$$dt = \left(\frac{R}{\sqrt{2gR}}\right) \frac{d\theta}{\sqrt{1 - \cos(\theta)}} = \sqrt{\frac{R}{2g}} \frac{d\theta}{\sqrt{1 - \cos(\theta)}},$$

then, integrating,

$$\int\limits_0^{t_P} dt = t_P = \sqrt{\frac{R}{2g}} \int\limits_0^{\cos^{-1}\left(\frac{2}{3}\right)} \frac{d\theta}{\sqrt{1 - \cos(\theta)}} = \infty,$$

as we showed in the previous problem. So, there's our puzzle. How can the sliding mass *ever* fly off the hemisphere at our precisely calculated values of θ_P and v_P if it takes an *infinitely long time* for the mass to get to point P? What's going on?

The answer is that our analysis has *assumed* that the point mass is actually sliding. But is it? At $\theta = 0$ (the top of the hemisphere) there is *zero tangential* gravitational force on the mass, and since the mass is given as motionless at $t = 0$, there is nothing to *start* the slide. The mass is initially in an equilibrium state at $\theta = 0$, $t = 0$, and there it will just sit, *forever*, which is exactly what the divergent integral for t_P is telling us. Now, of course, that equilibrium state is an *unstable* one (Problem 19 will have a more technical discussion of the *stability* of an equilibrium), and in the real world even the most minor disturbance would start the slide. Any puff of wind, the mere beating of the wings of a passing butterfly, the slightest flutter of an eyelash from a nearby observer—any of those would do the job.

In the perfect, blemish-free world of our theoretical mathematical analysis, however, there are no such disturbances. To have a slide starting at $\theta = 0$ we must *explicitly* introduce a disturbance, such as an initial *nonzero* speed for the mass at $\theta = 0$. That is, let's write

$$v(t = 0) = v(\theta = 0) = v_0 > 0.$$

This is the point that often goes unmentioned in textbooks.[1] The value of v_0 can be arbitrarily small, as *close* to zero as you want, but it cannot *be* zero.

To account for $v_0 > 0$, the conservation of energy equation we wrote earlier (see the second boxed equation) needs to be modified to read (the first term on the right is the kinetic energy of the initial disturbance)

$$\frac{1}{2}mv^2 = \frac{1}{2}mv_0^2 + mgR[1 - \cos(\theta)],$$

which becomes

$$v^2 = v_0^2 + 2gR[1 - \cos(\theta)].$$

So, if the mass departs the hemisphere at $\theta = \theta_P$, with speed v_P, we have (from the first boxed equation)

$$\cos(\theta_P) = \frac{v_0^2 + 2gR[1 - \cos(\theta)]}{gR} = \frac{v_0^2}{gR} + 2 - 2\cos(\theta_P),$$

and so

$$\cos(\theta_P) = \frac{2}{3} + \frac{v_0^2}{3gR}.$$

Notice that now, with $v_0 > 0$, θ_P is no longer independent of either g or of R (but, as before with $v_0 = 0$, the value of the mass still does not matter). Notice, too, that as v_0 *increases*, the value of θ_P *decreases*, which makes physical sense. Since the cosine function cannot exceed 1 for

---

[1] There *are* some exceptions. For example, in the extended fifth edition of the popular college freshman physics text by Halliday and Resnik, *Fundamentals of Physics*, Wiley, 1998, p. 180, we read that the mass is given a "very small push" to get it started. There is, however, no explicit discussion for *why* that is done (to achieve a finite sliding time), and there is no quantitative connection between that time and the magnitude of a "very small push." In the extended seventh edition by Halliday, Resnik, and Walker 2005, p. 192, the initial condition has been rephrased to be a "negligible" speed. That, I object, is not really correct. Small it may be, but if you want a finite t_P, then the initial speed cannot be called "negligible." Having $v_0 \neq 0$ literally makes all the difference in the world.

any real-valued angle, then

$$\frac{v_0^2}{3gR} \le \frac{1}{3},$$

which says that the initial speed of the mass must be in the interval $0 \le v_0 \le \sqrt{gR}$. What this means, physically, is that for any $v_0 > \sqrt{gR}$ the mass will shoot straight off the hemisphere, *with no sliding at all*. For $R = 3$ meters and $g = 9.81$ meters per second-squared, for example, the upper limit on v_0 is 5.42 meters per second.

Now, assuming $v_0 \le \sqrt{gR}$, the mass flies off the hemisphere (at $\theta = \theta_P$) with speed

$$v_P = \sqrt{v_0^2 + 2gR[1 - \cos(\theta_P)]} = \sqrt{v_0^2 + 2gR\left[\frac{1}{3} - \frac{v_0^2}{3gR}\right]},$$

or

$$v_P = \frac{1}{3}\sqrt{3}\sqrt{v_0^2 + 2gR}.$$

With all this information in hand, we can now calculate physically meaningful values for t_P. Instead of our divergent integral, we have

$$t_P = R\int_0^{\theta_P} \frac{d\theta}{v} = R\int_0^{\theta_P} \frac{d\theta}{\sqrt{v_0^2 + 2gR[1 - \cos(\theta)]}},$$

or

$$t_P = \sqrt{\frac{R}{2g}} \int_0^{\cos^{-1}\left(\frac{2}{3} + \frac{v_0^2}{3gR}\right)} \frac{d\theta}{\sqrt{\left(1 + \frac{v_0^2}{2gR}\right) - \cos(\theta)}}.$$

Before I give you your assignment, let me tell you what triggered my curiosity about t_P. In 1988 an elegant, *exact* analysis appeared for the problem of sliding on a hemisphere, an analysis that treats *any* value of μ (the coefficient of kinetic friction), and not just the friction-free case of $\mu = 0$.[2] Along with calculating the launch angle θ_P as a function of μ (for $\mu = 0$ the general result correctly reduces to the $48.2°$ we calculated earlier), the authors also calculated t_P. Their reported result, for the case of $\mu = 0$ (for a hemisphere with $R = 3$ meters in Earth's gravity), is 4.934 seconds with an initial speed of $v_0 = 0$ even though—as we've seen—the value of t_P for that particular case is actually *infinity*! To be blunt, Herreman and Pottel's t_P value for $\mu = 0$ simply cannot be correct.

To determine what nonzero v_0 is required for Herreman and Pottel's t_P, the MATLAB code **slide.m** evaluates the integral for t_P given in the following box. The code asks for keyboard-input values of v_0 in the first statement after the *while* command. Other than that, the code should be self-explanatory, with the exception (perhaps) of the line starting with *fun* (which stands for *function*, defining the integral's integrand). And, as you can probably easily guess, *acos* means arccosine or $\cos^{-1}$.

```
%slide.m/created by PJNahin for Mathematical Physics
R=3;g=9.81;Z=1;
F=sqrt(R/(2*g));
while Z>0
    v0=input('What is the initial speed?');
    B=1+(v0^2)/(2*g*R);
    fun=@(x) 1./sqrt(B-cos(x));
    A=(2/3)+(v0^2)/(3*g*R);
    A=acos(A);
    tp=F*integral(fun,0,A)
end
```

[2]W. Herreman and H. Pottel, "The Sliding of a Mass Down the Surface of a Solid Sphere," *American Journal of Physics*, April 1988, pp. 351, 372. Later, other writers repeated this calculation with some interesting variants of their own, with the most recent being (as I write) a paper by Felipe González-Cataldo, et al., "Sliding Down an Arbitrary Curve in the Presence of Friction," *American Journal of Physics*, February 2017, pp. 108–114. However, that paper (and all the earlier ones it cites) fails to address, or even mention, the issue of the duration of the slide.

To approximate the time reported in the paper by Herreman and Pottel for the $\mu = 0$ case, the code produced $t_P = 4.9501$ seconds for $v_0 = 0.0012$ meters per second and $t_P = 4.8267$ seconds for $v_0 = 0.0015$ meters per second. So, to achieve their $t_P = 4.934$ seconds an initial v_0 somewhere between 1.2×10^{-3} meters per second and 1.5×10^{-3} meters per second would be required. *Small*, yes, but *not* zero.

Now, for your assignment. Using **slide.m** as a guide, write code (in your favorite language) that, for $v_0 = 10^{-4}$ meters per second on a frictionless hemisphere with $R = 3$ meters, calculates t_P. How does the result change for $v_0 = 10^{-5}$ meters per second? Repeat both calculations for $R = 6$ meters.

PROBLEM 15
When the Computer Fails: When *Every* Day Is a Birthday

The previous discussions may have given the impression that armed with a computer, we can tackle even the toughest physics problem and just "simulate and compute our way to a solution." This discussion is to disabuse you of that happy (alas, false) view by showing you an easy-to-understand problem that so rapidly gets out of numerical control that mathematical *analysis* is the *only* option. Let's return, then, to the question of putting r distinguishable balls into n distinguishable boxes. In particular, how many distinguishable ways are there *that leave no box empty*? Obviously, this requires $r \geq n$, and if r and n are "reasonably small" then we can just enumerate all the ways. For example, if $r = 3$ balls and $n = 2$ boxes, the answer is 6, as shown in Figure P15.1.[1]

But what if $r = 5$ balls and $n = 3$ boxes? The answer is then 150 (and, I assure you, I did *not* enumerate all those ways!).

What I did was *iteratively calculate* the answer. That is, suppose we denote the answer by $A(r, n)$, where A stands for *arrangements* (that is, distinguishable distributions of the r distinguishable balls into the n distinguishable boxes). Now, imagine those n boxes all lined up,

---

[1] This is without Pauli exclusion. If exclusion is in effect, then the answer is clearly zero.

1. $\{ab|c\}$ **2.** $\{c|ab\}$

3. $\{a|bc\}$ **4.** $\{bc|a\}$

5. $\{ac|b\}$ **6.** $\{b|ac\}$

Figure P15.1. The six solutions.

from left (box 1) to right (box n). Since all the boxes must have at least one ball, let's first imagine that we put k balls into box n, where $1 \leq k \leq r$. (You might be wondering why k can run up all the way to r, the entire number of balls, which would leave no balls for the rest of the boxes. You'll see why this is okay in just a moment.) We can select those k balls in $\binom{r}{k}$ ways, and putting those k balls into box n leaves $r - k$ balls to be put into the remaining $n - 1$ boxes. Since $r - k$ balls in $n - 1$ boxes have $A(r - k, n - 1)$ arrangements, then the total number of arrangements of the k balls we picked for box n, combined with $A(r - k, n - 1)$ arrangements, gives a total of $\binom{r}{k} A(r - k, n - 1)$ arrangements for the case of k balls into box n. Since $1 \leq k \leq r$, then we simply sum over all possible k to get

$$
A(r, n) = \sum_{k=1}^{r} \binom{r}{k} A(r - k, n - 1).
$$

Now, before we calculate anything, we can make some general statements about $A(r, n)$, namely,

(a) $A(p, m) = 0,\ p < m;$[2]
(b) $A(p, 1) = 1,\ p \geq 1;$
(c) $A(p, p) = p!,\ p \geq 1.$

Statement (a) makes the physical observation that if we have more boxes (m) than balls (p), then it's impossible to put at least one ball in each box. Statement (b) says that if we have just one box, then

---

[2]This takes care of the concern I mentioned earlier about putting too many balls in box n and so not leaving enough balls for the other boxes.

there is just one arrangement (all the balls have to go into that box). And statement (c) we have already discussed in Problem 11 (the $n!$ arrangements of one ball in each of n boxes). So, for example, let's calculate $A(3,2)$, the number of distinguishable ways to put $r = 3$ distinguishable balls into $n = 2$ distinguishable boxes:

$$A(3,2) = \sum_{k=1}^{3} \binom{3}{k} A(3-k, 1)$$

$$= \binom{3}{1} A(2,1) + \binom{3}{2} A(1,1) + \binom{3}{3} A(0,1)$$

$$= 3A(2,1) + 3A(1,1) + A(0,1).$$

From (a) we have $A(0,1) = 0$, from either (b) or (c) we have $A(1,1) = 1$, and from (b) we have $A(2, 1) = 1$. Thus,

$$A(3,2) = 3(1) + 3(1) + 0 = 3 + 3 + 0 = 6,$$

just as shown in Figure P15.1 by direct enumeration.[3]

Now, with some additional mathematical manipulations (which I'll not do here) it can be shown that the boxed expression for $A(r, n)$ can be put into the following form, which

theoretically can be *directly* rather than iteratively calculated:[4]

$$\boxed{A(r, n) = \sum_{k=0}^{n} (-1)^k \binom{n}{k} (n - k)^r.}$$

---

[3]You should now check my iterative calculation of $A(5,3) = 150$.

[4]You can find an outline of how to do these manipulations in the classic book by William Feller, *An Introduction to Probability Theory and Its Applications* (3rd ed.), John Wiley & Sons, 1968, p. 60. See also pp. 101–102. Feller (1906–1970) was a professor of mathematics at Princeton, and his beautiful book should be on the shelf of all mathematical physicists.

You might be wondering why I write *theoretically*. After all, the second boxed calculation does seem to go through rather nicely:

$$A(3,2) = \sum_{k=0}^{2}(-1)^k \binom{2}{k}(2-k)^3 = \binom{2}{0}(2)^3 - \binom{2}{1}(1)^3 + \binom{2}{2}(0)^3$$
$$= 1(8) - 2(1) + 1(0) = 8 - 2 = 6,$$

just as we found before. What, you ask, is so theoretical about *that?*

Well, for small r and n all *is* okay,[5] but for large r and n the numbers that appear are simply enormous. (Try $r = 7$ and $n = 25$, values that actually aren't terribly large.) As Feller wrote in his book, the expression in the second box "provides a theoretical solution to an old problem but obviously it would be a thankless task to use it for the calculation of the probability … that in a village of $r = 1900$ people every day of the year is a birthday."[6] That is, if you set $r = 1,900$ and $n = 365$, then you will (metaphorically) be blown away by the blast wave of an arithmetic mega-explosion. There is, however, an elegant way out of this quandary, a way (that every mathematical physicist should have in his or her bag of tricks) that will let us easily calculate the probability of Feller's birthday problem.

The probability of there being no empty boxes when r distinguishable balls are placed in n distinguishable boxes is

$$P = \frac{A(r,n)}{n^r} = \sum_{k=0}^{n}(-1)^k \binom{n}{k}\frac{(n-k)^r}{n^r},$$

or

$$\boxed{P = \sum_{k=0}^{n}(-1)^k \binom{n}{k}\left(1 - \frac{k}{n}\right)^r.} \qquad (A)$$

What follows is how to find upper and lower bounds on P.

---

[5] Calculate $A(5,3)$ again, using the expression in the second box, and so confirm that $A(5,3) = 150$.

[6] You'll recall that in Problem 11 we calculated the probability that 365 people would result in a birthday every day of the year (365 is clearly the minimum number of people that could do that, but only with a *very* small probability—which we calculated). With 1,900 people the probability should be greater. But how *much* greater is our question now.

We start by writing, *for a given positive value of k*, the clearly true

$$(n-k)^k < (n)(n-1)(n-2)\ldots(n-k+1) < n^k,$$

and so

$$n^k \left(1 - \frac{k}{n}\right)^k < \frac{n!}{(n-k)!} < n^k,$$

and so

$$n^k \left(1 - \frac{k}{n}\right)^k < k! \binom{n}{k} < n^k,$$

and so

$$n^k \left(1 - \frac{k}{n}\right)^k \left(1 - \frac{k}{n}\right)^r < k! \binom{n}{k} \left(1 - \frac{k}{n}\right)^r < n^k \left(1 - \frac{k}{n}\right)^r,$$

or

$$\boxed{n^k \left(1 - \frac{k}{n}\right)^{k+r} < k! \binom{n}{k} \left(1 - \frac{k}{n}\right)^r < n^k \left(1 - \frac{k}{n}\right)^r.} \tag{B}$$

Next, recall the *logarithmic inequality* we derived in the preface:

$$\frac{x}{1+x} < \ln(1+x) < x, \qquad x > -1.$$

If we let $x = -\frac{k}{n}$ (and so $x > -1$, because the summation variable k in boxed expression (A) for P runs only up to n, and we ignore the case of $k = n$ because that term in the sum is zero), then we have

$$\frac{-\frac{k}{n}}{1 - \frac{k}{n}} < \ln\left(1 - \frac{k}{n}\right) < -\frac{k}{n},$$

or

$$-\frac{k}{n-k} < \ln\left(1 - \frac{k}{n}\right) < -\frac{k}{n}. \tag{C}$$

Consider, first, the upper-bound half of this double inequality:

$$\ln\left(1 - \frac{k}{n}\right) < -\frac{k}{n},$$

which says

$$r \ln\left(1 - \frac{k}{n}\right) < -k\frac{r}{n},$$

which says

$$\ln\left(1 - \frac{k}{n}\right)^r < -k\frac{r}{n},$$

which says

$$\left(1 - \frac{k}{n}\right)^r < e^{-k\frac{r}{n}},$$

which says

$$n^k\left(1 - \frac{k}{n}\right)^r < n^k e^{-k\frac{r}{n}} = \left(ne^{-\frac{r}{n}}\right)^k. \tag{D}$$

Next, consider the lower-bound half of the double inequality in (C):

$$-\frac{k}{n-k} < \ln\left(1 - \frac{k}{n}\right),$$

which says

$$-\frac{k}{n-k}(k+r) < (k+r)\ln\left(1-\frac{k}{n}\right) = \ln\left(1-\frac{k}{n}\right)^{(k+r)},$$

which says

$$e^{-k\frac{k+r}{n-k}} < \left(1-\frac{k}{n}\right)^{(k+r)},$$

which says

$$n^k e^{-k\frac{k+r}{n-k}} < n^k\left(1-\frac{k}{n}\right)^{(k+r)},$$

or

$$\left(ne^{-\frac{k+r}{n-k}}\right)^k < n^k\left(1-\frac{k}{n}\right)^{(k+r)},$$

or, finally, using (B),

$$\left(ne^{-\frac{k+r}{n-k}}\right)^k < k!\binom{n}{k}\left(1-\frac{k}{n}\right)^r. \tag{E}$$

Combining the results of (B), (D), and (E),

$$\left(ne^{-\frac{k+r}{n-k}}\right)^k < k!\binom{n}{k}\left(1-\frac{k}{n}\right)^r < \left(ne^{-\frac{r}{n}}\right)^k, \tag{F}$$

where the upper and lower bounds in (B) have been replaced with even weaker (less restrictive) bounds. The reason all this has been useful is that *for any given value of* k, the ratio of the upper and lower bounds in

(F) approaches 1 as $n \to \infty$. That is,

$$\lim_{n \to \infty} \frac{\left(ne^{-\frac{r}{n}}\right)^k}{\left(ne^{-\frac{k+r}{n-k}}\right)^k} = \lim_{n \to \infty} \frac{e^{-k\frac{r}{n}}}{e^{-k\frac{k+r}{n-k}}} = \frac{1}{1} = 1.$$

Even with the weaker bounds, the interval in which $k! \binom{n}{k} \left(1 - \frac{k}{n}\right)^r$ exists is being *squeezed*, becoming (in a relative sense) narrower and narrower as $n \to \infty$, and so (speaking intuitively, now) it *doesn't matter* what particular value we assign to $k! \binom{n}{k} \left(1 - \frac{k}{n}\right)^r$ as long as it is in that ever-shrinking interval. In fact, let's use the upper limit in (F) and define a parameter λ as

$$\lambda = ne^{-\frac{r}{n}},$$

and then assume that both n and r become ever larger in such a way that λ is always finite. That allows us to assign the value λ^k to $k! \binom{n}{k} \left(1 - \frac{k}{n}\right)^r$, for each value of k, with little resultant error.[7] That is, for "large" r and n, we have the so-called Poisson approximation:[8]

$$\binom{n}{k} \left(1 - \frac{k}{n}\right)^r \approx \frac{\lambda^k}{k!}.$$

So, returning to (A), the probability that there are no empty boxes when r distinguishable balls are distributed at random in n distinguishable boxes is given by, in the limit as $r, n \to \infty$,

$$P = \sum_{k=0}^{\infty} (-1)^k \binom{n}{k} \left(1 - \frac{k}{n}\right)^r = \sum_{k=0}^{\infty} (-1)^k \frac{\lambda^k}{k!}.$$

This last sum will, in an instant, be recognized by any mathematical physicist (and also by any freshman calculus student who has been

---

[7] Mathematicians would prefer that I *prove* this claim. The argument I've just taken you through is what most mathematical physicists would accept, however, and so this is a good example of the difference between physicists and mathematicians.

[8] After the French mathematical physicist Siméon Denis Poisson (1781–1840).

paying attention in class) as the power series expansion of $e^{-\lambda}$.[9] And so, just like that, we have the probability of putting r distinguishable balls into n distinguishable boxes so that no box is empty:

$$P = e^{-ne^{-\frac{r}{n}}},\qquad\text{(G)}$$

a "stacked" exponential.

Your assignment is to use (G) to determine when P first exceeds 0.5 in the "every day of the year a birthday" question. That is, how large a group of people is required for the probability to first exceed $\frac{1}{2}$ that every day of the year is a birthday. *Hint*: Set $n = 365$.

Problem 16
When Intuition Fails: Sometimes What *Feels* Right, Just Isn't

In the previous mathematical problem, your intuition was almost surely challenged, at least a little. That's the particular charm, I think of the birthday problem (in all its variations). In physics, as well, such counterintuitive situations are not at all difficult to find, with quantum mechanics being the most obvious branch of physics in which just about *everybody's* intuition gets trampled. To quote Richard Feynman, "One [has] to lose one's common sense in order to perceive what [is] happening at the atomic level."[1] And then later, in the same work (p. 10), "The theory of quantum electrodynamics describes Nature as absurd from the point of view of common sense." Twenty years earlier, in his November 1964 Messenger Lectures at Cornell University, Feynman had been equally blunt: "I think I can safely say that nobody understands quantum mechanics."[2]

---

[9] Refer to the end of Problem 6.

[1] From Feynman's book *QED: The Strange Theory of Light and Matter*, Princeton University Press, 1985, p. 5. *QED* is based on lectures Feynman gave in July/August 1979 at the University of Auckland, New Zealand, and again in May 1983 at UCLA, on quantum electrodynamics (the theory on how photons and electrons interact).

[2] See Feynman's *The Character of Physical Law*, MIT Press, 1967, p. 77.

We don't have to jump into quantum physics, however, to find a physical situation that is even more nonintuitive than the birthday problems. Imagine a box with n balls in it, identical in every respect except for being numbered from 1 through n (that is, the balls are distinguishable). Further imagine that you draw balls from the box, one at a time, first recording the number of each drawn ball and then replacing the ball in the box. After each replacement you give the box a vigorous and prolonged shaking to thoroughly mix up the balls before the next drawing.

It's pretty obvious (I hope) that after $n+1$ drawings you are *certain* to have drawn a ball that you drew earlier. But of course you *could* draw a previously drawn ball much earlier than that, as early, in fact, as the *second* drawing (the first drawing after the initial drawing). If n is a really large number, however, it may seem just as obvious that to get a repeat drawing *early* in the sequence of drawings would be an unlikely event. When I was teaching introductory probability theory to undergraduate electrical engineering students, I would ask them what their intuition suggested and, routinely, I'd get guesses like $\frac{1}{2}n$. That is, for a box with 20,000 balls, the intuitive answer was, typically, that it would take something on the order of 10,000 draws before a repeat would be likely (where just what "likely" means was left initially undefined; one possibility is to ask for the greatest number of drawings for which the probability of not having a repetition remains greater than $\frac{1}{2}$).

My students were always astonished when I told them, "No, that's far too large. If you have a lot of people, each with his or her own box containing 20,000 balls, you'll find that *on average* it takes fewer than 180 draws to get a repeat." They didn't believe me until I showed them how to derive a formula for the average number of draws. (Some didn't believe me even *after* that derivation, but I'll tell you more about *that* in just a bit.)

Now, before deriving that formula for the average number of draws after the first draw, until we get a repeat, let me motivate for you a central result from probability theory. Suppose you are observing the voltage v of an electrical wire, a voltage that can take on (at random) any one of S values. These values are $v_1, v_2, \ldots, v_S$. Each time you measure v you will get one of those values. Suppose you make M measurements and record v_1 a total of m_1 times, v_2 a total of m_2 times, and so on, up to recording v_S a total of m_S times. Of course, $m_1 + m_2 + \ldots + m_S = M$. The *average value* of v (often called the *expected value*) is written as $E(v)$. From grade school arithmetic we have

$$E(v) = \frac{v_1 m_1 + v_2 m_2 + \ldots + v_S m_S}{M} = v_1 \frac{m_1}{M} + v_2 \frac{m_2}{M} + \ldots + v_S \frac{m_S}{M}$$

$$= v_1 \text{Prob}(v = v_1) + v_2 \text{Prob}(v = v_2) + \ldots + v_S \text{Prob}(v = v_S),$$

or

$$E(v) = \sum_{k=1}^{S} v_k \text{Prob}(v = v_k).$$

So, if we define N as the number of drawings (after the first drawing) from a box of n balls until we get a repeat, what we want to calculate is

$$E(N) = \sum_{k=1}^{n} k \text{Prob}(N = k).$$

Our task is now clear: we have to calculate the probabilities $\text{Prob}(N = k)$.

To do that, I'll work through the details of the first few values of k, and then I think you'll see the general answer. To start, what's the probability that $N = 1$; in other words, what's the probability that we immediately draw the same ball as we originally drew? On the original draw, we drew some ball (*any* of the n in the box), an event that happens with probability $\frac{n}{n}$. Then, on the first draw after that original draw we draw the same ball, which happens with probability $\frac{1}{n}$. Since the draws are independent because of the vigorous shaking of the box, we have

$$\text{Prob}(N = 1) = \left(\frac{n}{n}\right)\left(\frac{1}{n}\right) = \frac{1}{n}.$$

What's the probability $N = 2$? On the original draw we draw any ball (probability $\frac{n}{n}$), and on the first draw after the original draw we draw any other ball (probability $\frac{n-1}{n}$). Then, on the second draw after the original draw we draw *either one* of the two previously drawn balls (probability $\frac{2}{n}$). So,

$$\text{Prob}\,(N=2) = \left(\frac{n}{n}\right)\left(\frac{n-1}{n}\right)\left(\frac{2}{n}\right) = \frac{2\,(n-1)}{n^2}.$$

What's the probability $N = 3$? On the original draw we draw any ball (probability $\frac{n}{n}$), on the first draw after the original draw we draw any other ball (probability $\frac{n-1}{n}$), and on the second draw after the original draw we draw any yet-undrawn ball (probability $\frac{n-2}{n}$). Then, on the third draw after the original draw we draw any one of the *three* previously drawn balls (probability $\frac{3}{n}$). So,

$$\text{Prob}\,(N=3) = \left(\frac{n}{n}\right)\left(\frac{n-1}{n}\right)\left(\frac{n-2}{n}\right)\left(\frac{3}{n}\right) = \frac{3\,(n-1)\,(n-2)}{n^3}.$$

Do you see the pattern? Sure you do!

$$\text{Prob}\,(N=k) = \frac{k\,(n-1)\,(n-2)\ldots(n-k+1)}{n^k},$$

and so

$$E\,(N) = \sum_{k=1}^{n} k^2 \left(\frac{1}{n}\right)\left(\frac{n-1}{n}\right)\left(\frac{n-2}{n}\right)\ldots\left(\frac{n-k+1}{n}\right).$$

This expression is easily evaluated on a computer, even for large values of n, and the MATLAB code **balls.m** does the job.

```
%balls.m/created by PJNahin for Mathematical Physics
n=input('How many balls in the box?')
E=0;
for k=1:n
    F=1/n;
    for j=1:k-1
        F=F*(n-j)/n;
    end
```

(continued)

(continued)

```
F=k*k*F;
    E=E+F;
end
E
```

When run, the code produced the following results:

Number of balls in box *n*	Average number of draws until a repeat $E(N)$
10	3.66
100	12.21
1,000	39.3
10,000	125
20,000	177
50,000	280

If you change the statement *F=k*k*F;* to *F=k*F*; then **balls.m** computes $\sum_{k=1}^{n} \text{Prob}(N=k)$, which should equal 1 for *any* positive integer value of *n* (and not to keep you in suspense, it does!). Can you prove that *analytically*? Notice that $E(N)$ does not increase *linearly* with increasing *n* but, rather, grows more slowly. The numerical results from the code suggest, in fact, that the increase is proportional to some *fractional power* of *n*, which can be confirmed theoretically by calculations more advanced than I'm presenting in this book.

Challenge Problem 1: For my students who were still not quite convinced by all this, I suggested they write a Monte Carlo simulation. So, there's *your* first assignment: write a Monte Carlo code (in your favorite language) to simulate the drawing (and replacing) of balls from a box containing 1,000 balls until a repeat occurs, and compare the simulation's average number of drawings (after the original draw) with the theoretical value of 39.3.

Challenge Problem 2: Define the integer T to be the greatest number of drawings for which the probability of *not* having a repeat drawing is still more than $\frac{1}{2}$. That is, what is T if for $T + 1$ drawings the probability of *getting* a repeat first exceeds $\frac{1}{2}$.

Challenge Problem 3: Assuming $E(N) = kn^x$ (as suggested by the computer results of **balls.m**), where k and x are constants, estimate the values of k and x.

PROBLEM 17
Computer Simulation of the Physics of NASTYGLASS: Is This Serious? Maybe

Cautionary Note: The following is the most speculative presentation in the book. Some of what you read here may at first seem to be "way out," but at no place is there a purposeful misstatement of either the physics or the math. So, keep an open mind.

In this problem we'll be concerned with the idea of a *digital picture*, a once-exotic concept, but now one familiar to even grade-school children (many of whom have a smartphone equipped with a digital camera). In the traditional, printed *analog* picture (a photograph), the brightness level is thought of as varying smoothly from black to white (I'm limiting my discussion here to just black-and-white, or b/w, images) through all intermediate *gray levels* as we move smoothly about the picture along any continuous path.

A digital picture, in contrast, is created from an analog one by *triple quantization* (once in the horizontal direction and again in the vertical direction, creating so-called picture elements, or *pixels*, and for a third time on the gray level of each pixel). If there are h pixels across the horizontal extent of the analog picture, and v pixels across the vertical extent of the analog picture, then there are a total of hv pixels. If the gray level of each pixel is quantized to b bits, then there are 2^b discrete possible brightness levels for each pixel (usually, the value zero is assigned to the blackest black, and the value $2^b - 1$ is assigned to the whitest white).

These simple observations immediately lead us to the interesting conclusion that there are a *finite* number, N, of possible digital pictures:

$$N = \left(2^b\right)^{hv} = 2^{hbv}.$$

The totality of this collection (called The Universal Picture Album or TUPA[1]) may be finite, but that doesn't mean it's small. For example, suppose hv equals 10 million (that is, a b/w TUPA picture is 10 megapixels in size, comparable to what a good digital camera produces), and $b = 11$ bits (and so there are 2,048 gray levels); each TUPA picture would therefore be a *very* crisp, ultrasharp, high-resolution photo. For these assumed values of quantization, the number of pictures in TUPA is

$$N = 2^{110,000,000} \approx 10^{33,113,299}.$$

That's a 1 followed by a lot more than 33 million zeros![2] Each of these pictures is of course indistinguishable from a multitude of others (think of two pictures differing by just one analogous pixel in each picture being different by just one level of gray), and some are simply uninteresting (the ones, for example, with every pixel at the same gray level). But there are also a lot of quite interesting pictures.

Here are some of them. There is in TUPA a picture of every creature in the universe that has ever lived (including you), of every creature that ever *will* live, and of every creature that can only be imagined. There is in TUPA an image of every page from every book that has been or *will* be written, in every language that has existed or ever *will* exist. There is an image of every painting that has been or ever *will* be painted. In TUPA is the frame-by-frame imagery of every Hollywood movie that has been or ever *will* be made (in many of which *you*

---

[1] For the origin of TUPA, see my book *Number-Crunching: Taming Unruly Computational Problems from Mathematical Physics to Science Fiction*, Princeton University Press, 2011, pp. xvii–xx.

[2] Mathematicians have defined numbers far larger than N. With the *googol* defined to be 10^{100}, for example, the *googolplex* is 10^{googol}, that is, a 1 followed by a googol of zeros. This is immensely larger than N, but it lacks the fantastic *physical* interpretation of N.

appear), as well as all those movies that will never be made. Of great particular interest to physicists is the fact that somewhere in TUPA are the plans for a time machine, a matter transmitter, and a faster-than-light space drive. And mathematicians can only wonder at the TUPA pictures that reveal the solutions to all possible mathematical problems. On and on it goes—there is an image of *everything* that has existed or will or might (or might not) exist throughout the entire universe. Perhaps even more astonishing, TUPA contains itself as a subset, because in TUPA there are images of ever more slightly reduced versions of every image in TUPA![3]

Now, with all that as background, we at last come to the central question of this seemingly outlandish discussion: is it possible for a digital picture to be physically painful to look at? This question was first raised some years ago in the British publication *New Scientist*,[4] and is not as frivolous as it may at first seem. To be sure there is no misunderstanding, the question is not about pictures that evoke a distressful *emotional* reaction (a hungry child, for example) but, rather, about the possibility of pictures that might be so awful to view that they could cause neurological discomfort, such as one experiences by stepping on a tack with a bare foot.

Now, surely nearly everyone reading that will balk at this point, declaring that they have never seen such a picture. Well, neither have I, but keep in mind the absolutely gargantuan size of TUPA. Throughout all history, all the people who have ever lived have collectively seen only an infinitesimal fraction of TUPA; there are vastly more pictures in TUPA than there are elementary particles in the entire universe (estimated to be far less than a googol). So, can we *really* declare with authority that there are *no* pictures that can cause physical pain? Who knows *what* lurks in the unfathomable expanse of TUPA?

---

[3]At the risk of overly exciting the interest of WikiLeaks, it is amusing to point out that in TUPA there is a photograph of every page of every *Top Secret* Pentagon, CIA, FBI, and Kremlin document ever written or that *will be* written. Signed confessions from the villains behind the disappearances of Judge Crater and Jimmy Hoffa are in TUPA. Even all of President Trump's complete tax returns (including those under audit), and Hillary's e-mails, are in TUPA!

[4]David E. H. Jones, *New Scientist*, July 22, 1971, p. 222, and reprinted in Jones's book *The Inventions of Daedalus*, W. H. Freeman, 1982, pp. 158–159.

Now, one obvious way to try to answer that question is to randomly generate digital pictures and to simply look at them. If there is such a thing as a tack-like image in TUPA, then even though we cannot even begin to imagine what must be its strange characteristics, we can be confident that we will know instantly if we should happen to stumble upon it. It will *hurt*! This means such a picture will require no special skills to detect it. Anyone with the sense to say *ouch* could find it. Since there might well be a Nobel Prize for such a discovery, a little pain seems a small price to pay for immortality.

Alas, there is just one fundamental problem with that approach. Think of a vast (I mean, *really* vast) lawn of thick grass that extends for hundreds of thousands of millions of billions of trillions of light-years in all directions, with every grass shoot representing a TUPA picture. Each such shoot that represents a painful picture has been replaced with a tack. A random generation of pictures is analogous to running barefoot through this truly enormous grass field and hoping, by chance, to step on a tack. Not likely, I'd say.[5]

It was Jones (note 4) who suggested that physics[6] and computer technology might provide an alternative to this gloomy appraisal. Jones's key idea was motivated by the *aural* discomfort caused by what electrical engineers call *crossover distortion* in push-pull audio amplifiers. Such amplifiers (see any good book on electronics) use a *pair* of active devices (for example, vacuum tubes or transistors): with a sinusoidal input signal (a single frequency tone) each device operates during just half of the sine-wave cycle, with the other device inoperative (*cut off*, in engineering jargon). When the sine-wave amplifier input crosses through zero volts to begin the other half of its cycle, the operative device becomes inoperative, and the inoperative one becomes operative. In theory, this periodic swapping of roles (hence

---

[5]Suppose we could computer-generate one hundred thousand million billion trillion TUPA images per femtosecond (a billionth of a microsecond). That's 10^{47} images per second. The universe is estimated to be 15 billion years old, and so if our computer started its Herculean task at Big Bang time, it would have produced (to date) a total of $(1.5 \times 10^{10})(365)(24)(3600)(10^{47}) = 4.73 \times 10^{64}$ pictures. That may seem like a lot, but it's really pretty small potatoes compared with N.

[6]Dr. Jones (born 1938) is actually a chemist, not a physicist, but science knows no artificial boundaries (the laws of physics are the same for everybody).

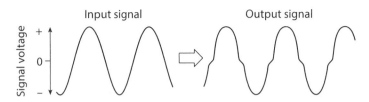

Figure P17.1. Crossover distortion in a push-pull audio amplifier.

the picturesque name as "push-pull") takes place instantly at the zero-crossing instants, but in practice there is usually an interval of input signal values for which *both* devices are cut off (or nearly so), and the gain of the amplifier is momentarily reduced. The output waveform such as shown in Figure P17.1 results, a waveform showing the savage ravages of crossover distortion.

So why are push-pull amplifiers *ever* used? The reason is that such amplifiers are energy efficient compared with single-device designs, in which the lone active device is *always* operative (that is, is drawing energy from the amplifier's power supply). In push-pull, however, when neither device is operative during crossover (when there is no input signal) the demands on the power supply are greatly reduced. Such intervals of quiet are common in broadcast radio (excluding the more hysterical political talk-radio programs), and so the push-pull amplifier is most often found in applications with a limited available supply of energy, such as in portable, battery-powered radios. There *is*, however, a price paid for this efficiency. It is a well-known empirical observation that even seemingly small amounts of crossover distortion can result in a perfectly awful sounding amplifier output.

With that in mind, Jones asked himself a *what if* question. What if, he wondered, one made a glass window that *intentionally* introduced crossover distortion into the light (a sinusoidal waveform) incident upon it? What would things look like after the light it scattered passed through such a window? Since *audio* crossover distortion *sounds* terrible, then is it perhaps reasonable to speculate that *video* crossover distortion would *look* terrible—so terrible, in fact, as to hurt any who gazed upon it? Jones concluded that this would indeed be the effect of such a window, and he named it NASTYGLASS. He arrived at this conclusion, however, strictly on the basis of the audio analogy, admitting that he actually "cannot predict how things will look seen through NASTYGLASS."

That admission, in turn, sparked *my* imagination, as I realized that with the aid of a modern digital computer image-processing system we *can* determine the visual effect of such a glass, even though no one has (as far as I know) made a NASTYGLASS window.[7] Here, in fact, we have a really nifty example of an imaginary physics problem, something that I hinted at near the end of the preface. The key idea is that now we're not talking about a random search for a tack-like picture. Rather, now we have an understandable physical process (NASTYGLASS) with which to map a "nice" digital picture (more on what "nice" means in just a bit) into a new digital picture that suffers from video crossover distortion and so, by audio analogy, should be "less nice."[8] If we're lucky, the result will be an image that, as Jones speculated, will be so "not nice" as to have a "leering, sickly quality" to it, perhaps even one "so horrid that no Peeping Tom could bear to look" at it. That's certainly a high (or low, if you prefer) goal to shoot for, but nothing is worth pursuing if it is too easy. To paraphrase Einstein, "Anyone can pound a nail through a *thin* board. But a *thick* board, well, that's a different story!"

Okay, down to business. How would NASTYGLASS work if we *did* make it?

To start, imagine a beam of light consisting of just one wavelength (a pure "tone of light") incident on an *ordinary* glass window pane. The alternating electric field of the light induces alternating currents

[7]Jones did briefly mention the possibility of making NASTYGLASS from a so-called Ovshinsky amorphous semiconductor, named after the American inventor Stanford Ovshinsky (1922–2012). An amorphous solid has no crystalline structure (ordinary glass is an example), such as we commonly associate with semiconductor materials like silicon and germanium.

[8]The manipulation of gray levels is something photographers have been doing since the first blemish on a young maiden's nose was tastefully removed from some ancient daguerreotype. As is well known, the photography wizards at *Playboy* and its competitors have elevated analog manipulation of gray levels to a near art form. But with NASTYGLASS, and a computer simulation of it, we can replace the arbitrary, time-consuming nature of human manipulation with the rigor of mathematical physics automatically performed at electronic speed.

in the glass; the greater these currents are, the greater is the energy loss of the beam (and the greater is the increase in the temperature of the glass). The directly observable result is that the light beam emerges on the other side of the window, attenuated in brightness. A perfectly transparent (invisible) window would support *no* such energy-dissipating currents: it would be a perfect insulator. A window deviates ever more from the perfection of invisibility the more it becomes a conductor. A perfectly conducting window would be impossible to look through.[9]

For an incident electric field strength *below* a certain threshold (denoted by E_t), NASTYGLASS is a perfect conductor that completely "short-circuits" the light beam. Once the electric field strength exceeds this threshold, however, NASTYGLASS switches instantly to being a perfect insulator[10] and so transmits the light beam with no attenuation, until the incident electric field strength again falls below the threshold. This behavior is shown in Figure P17.2, which depicts the time variation of the electric field of a pure "tone of light" emerging from a NASTYGLASS window. As you can see, the waveform is displaying classic video crossover distortion.

To express NASTYGLASS in analytic form, concentrate your attention on a single (arbitrary) pixel of the digital picture that emerges from the NASTYGLASS window. As an *output* image pixel we'll denote its gray-level brightness by B_o. The corresponding pixel of the *input* image on the incident side of the window will have brightness B_i, and since the window is a dissipative medium, we must have $B_o \leq B_i$ (with

[9]As all who recall their high school physics know, the relationship between voltage, current, and resistance is given by Ohm's law (for small values of voltage and current). For the case of the time-varying high-frequency electric fields in visible light, the voltage and currents are also time-varying, high-frequency quantities, and Ohm's law is in the form of $r = \Delta v / \Delta vi$. Since Δv and Δi are *changes* in the incident voltage and the resulting current, r is called the *dynamic* resistance.

[10]As Figure P17.2 shows, once the electric field strength of the incident light becomes sufficiently large, then *changes* in the voltage (that is, Δv) result in $\Delta i = 0$. Thus, $r = \infty$ and NASTYGLASS has infinite (or, at least, a very large) dynamic resistance. Otherwise, the dynamic resistance is "zero" (that is, a "small" Δv results in a "large" Δi).

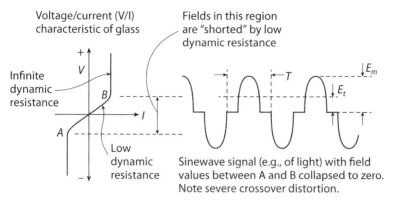

Figure P17.2. The physics of NASTYGLASS.

equality occurring if the NASTYGLASS electric field switching thresh-old E_t is zero). Our question now is, what is the precise relationship between B_o and B_i?

The perceived brightness (that is, gray level) of a pixel is a measure of the *power* level of the light radiating from that pixel. The power, in turn, is directly related to the average value of the squared electric field of the light. Therefore, if k is a scaling factor that makes the units come out right, then

$$B_i = \frac{k}{T} \int_0^T E_i^2(t)\, dt,$$

and

$$B_o = \frac{k}{T} \int_0^T E_o^2(t)\, dt,$$

where, as shown in Figure P17.2, T is the *half*-period of the light. If we write the incident electric field as

$$E_i(t) = E_m \sin\left(\frac{\pi t}{T}\right),\ 0 \le t \le T,$$

then, since there is no output electric field until $E_i(t) > E_t$, the output electric field is zero until that condition is satisfied, and then it is given by

$$E_o(t) = E_m \sin\left(\frac{\pi t}{T}\right), \quad \frac{T}{\pi}\sin^{-1}\left(\frac{E_t}{E_m}\right) < t < T\left[1 - \frac{1}{\pi}\sin^{-1}\left(\frac{E_t}{E_m}\right)\right].$$

Thus,

$$B_i = E_m^2 \frac{k}{T}\int_0^T \sin^2\left(\frac{\pi t}{T}\right) dt = \frac{kE_m^2}{2},$$

while if $E_t < E_m$,

$$\begin{aligned}
B_o &= E_m^2 \frac{k}{T}\int_{\frac{T}{\pi}\sin^{-1}\left(\frac{E_t}{E_m}\right)}^{T\left[1-\frac{1}{\pi}\sin^{-1}\left(\frac{E_t}{E_m}\right)\right]} \sin^2\left(\frac{\pi t}{T}\right) dt \\
&= E_m^2 \frac{k}{\pi}\left[\frac{\pi}{2} - \sin^{-1}\left(\frac{E_t}{E_m}\right) + \frac{1}{2}\sin\left\{2\sin^{-1}\left(\frac{E_t}{E_m}\right)\right\}\right].
\end{aligned}$$

That is, $B_o = 0$ if $E_m < E_t$, otherwise

$$\boxed{B_o = \frac{2B_i}{\pi}\left[\frac{\pi}{2} - \sin^{-1}\left(\frac{E_t}{E_m}\right) + \frac{1}{2}\sin\left\{2\sin^{-1}\left(\frac{E_t}{E_m}\right)\right\}\right].}$$

Notice that the boxed expression says $B_o = 0$ when $E_t = E_m$ (which is physically correct), and that $B_o = B_i$ when $E_t = 0$ (which is physically correct).

Now, to sidestep the question of just what are the electric fields E_t (a property of NASTYGLASS) and E_m (the strength of the incident light), we can express the boxed equation in a more convenient way. Suppose we write B_L to denote the least incident image pixel gray level that will just produce an output on the other side of the window. The value of B_L, varying from 0 to $2^b - 1$ (in integer steps) for b bits of gray-level quantization, determines how nasty our NASTYGLASS window is: $B_L = 0$ means it isn't nasty at all, while $B_L = 2^b - 1$ means it is a totally nasty window. Whatever the value of B_L, it is associated with an incident light intensity having E_t as its *maximum* electric field.

Thus,

$$B_L = k\frac{E_t^2}{2},$$

or

$$E_t = \sqrt{\frac{2B_L}{k}}.$$

Also,

$$B_i = k\frac{E_m^2}{2},$$

and so

$$E_m = \sqrt{\frac{2B_i}{k}}.$$

So, finally, we have the answer to our question of how the gray-level changes for each pixel as an image transits a pane of NASTYGLASS:

$$B_o = B_i\left[1 - \frac{2}{\pi}\sin^{-1}\left(\sqrt{\frac{B_L}{B_i}}\right) + \frac{1}{\pi}\sin\left\{2\sin^{-1}\left(\sqrt{\frac{B_L}{B_i}}\right)\right\}\right], \quad B_i \geq B_L$$

$$B_o = 0, \; B_i \leq B_L$$

This result[11] tells us precisely how NASTYGLASS maps an incident pixel gray level B_i into an output pixel gray level, with a single degree

---

[11]Notice that if the window is not nasty at all, that is, if $B_L = 0$, our result says $B_o = B_i$. In other words, every input image pixel gray level is transmitted *without alteration*, which is just what we'd expect. However, for a totally nasty window, in which B_L is so large that $B_i \leq B_L$ for every input pixel, then $B_o = 0$ for every output pixel. To put it bluntly, in that case the output image would be, truly and in the most physical sense, an image from the "dark side."

of freedom represented by a characteristic *nasty transmission parameter* B_L. As befits a window dubbed "nasty," the preceding boxed expression is a nonlinear transformation sufficiently convoluted to be properly called pretty nasty in its own right (although plots of it generate a family of deceptively benign, smooth-looking curves).

We can now use this analytic result to computer-simulate NASTY-GLASS. Indeed, 30 years ago, when I first came across the concept of NASTYGLASS, I ran some simulations at the minicomputer laboratory operated by the University of New Hampshire's Electrical Engineering Department.[12] For the input image, I used an analog photograph of one of the great men of science, Albert Einstein. In addition to holding a lofty position in the world of physics, he was also well known for his humanitarianism and gentle spirit. So, selecting one of his commonly available "grandfatherly" images, I felt sure that if NASTYGLASS really *is* nasty, then its beastly effect on such a heartwarming picture would be easily apparent in the computer's output image. Placed in front of a TV camera/digitizer, the resulting digital picture was then input to a computer programmed with the NASTYGLASS pixel mapping function, and output images were displayed on a TV monitor.

You can find photographs of those output images in my 1987 paper (note 12), but I haven't reproduced them here because, owing to an unfortunate computational error, I used an incorrect pixel mapping function. (Also, I seem to have misplaced them.) Still, that erroneous mapping was close enough to being correct that as B_L was increased, Einstein's output image *did* seem to take on an increasingly more menacing appearance—not a pretty sight, by any means, but *not* physically painful to look at, either. Perhaps with the correct pixel mapping function derived here the results would be different. I'll leave that to an adventurous reader to explore further. Please let me know if you do take up this challenge!

Now, here's a more elementary challenge for you to consider. A tremendous advantage that digital pictures have over analog pictures is

---

[12]You can find a report of that work in my paper "Looking at the World Through a NASTYGLASS Window," *IEEE Aerospace and Electronic Systems Society Magazine*, December 1987, pp. 2–7.

the ease with which they can be electronically manipulated. The gray level of any pixel can be mapped into any other gray level through the use of a simple table look-up process. Suppose the gray level of a pixel at coordinates (x, y) in an input image is g_i, where $0 \leq g_i \leq 2^b - 1$ for b bits of quantization. We then imagine that we have a table in a computer memory with 2^b entries, and that if we go to entry g_i in that table, we'll find the number g_o $(0 \leq g_o \leq 2^b - 1)$, the gray level to be assigned to the pixel at (x, y) in the output image. That is, $g_i \rightarrow g_o$ is a *mapping* for the gray level of each pixel in the input image to the corresponding output image pixel gray level. To be specific, suppose entry g_i in the table is $2^b - 1 - g_i$. That is, $g_i = 0$ maps into the gray level $2^b - 1$, and $g_i = 2^b - 1$ maps into the gray level 0, to treat the two extreme cases. The resulting output image is called the *negative* of the input image, which of course, has a well-known analog equivalent, achieved with messy chemicals.

But now consider this clever twist. Suppose entry g_i in the look-up table is g_i for $0 \leq g_i \leq 2^{b-1}$ (that is, the first half of the table), and $2^b - 1 - g_i$ for the second half. The resulting output image is called *half-reversed*, as the darker portions of the input image are unchanged, while the brighter portions, *only*, are folded back into the darker gray levels. Such a modified picture, odd looking though it may be, can often render visible details in the bright areas that are otherwise lost to the eye (which is less able to detect brightness level *changes* as the brightness increases). There is, as far as I know, no equivalent technique in analog photography.

Since there are 2^b different gray levels with b bits of quantization, and since we can load each of the 2^b elements of the look-up table with any number from 0 to $2^b - 1$ (one can put the same number into more than one element of the look-up table), there are $(2^b)^{(2^b)}$ possible look-up tables. Evaluate this number for $b = 11$, and write it in scientific notation. That is, as $p \times 10^q$, where q is an integer and p is a number between 1 and 10. *Hint*: It's far larger than a googol, but far less than a googolplex.

Here are two more number crunchers for you. First, when I wrote earlier about the vastness of TUPA, I compared it to a lawn of dense grass shoots that "extends for hundreds of thousands of millions of billions of trillions of light-years in all directions." Was that mere hyperbole? Or was it instead a gross *under*estimate? To answer that question, suppose a lawn has 100 grass shoots per square inch. What would be the radius of a circular lawn that contains N grass shoots? Use 186,210 miles per second as the speed of light. Also, suppose each 10-megapixel TUPA image is printed out as an $8'' \times 10''$ glossy. If each pixel is a square, how many pixels are there per horizontal/vertical inch in the image?

PROBLEM 18

The Falling Raindrop, Variable-Mass Problem: Falling *Slower* Than Gravity

Imagine a spherical water droplet, with initial radius $a \geq 0$ at time $t = 0$, beginning to fall under gravity from rest (that is, its downward speed at $t = 0$ is $v = 0$) through a uniformly dense fog. That is, through an atmosphere saturated with water vapor.[1] At any time $t \geq 0$ we'll write the mass of the droplet as $m(t)$, with radius $r(t)$ (and so $r(0) = a$). If we assume the mass of the droplet increases in direct proportion to its surface area (this is called a *mass accretion rule*, and it's not the only possibility; we'll explore other possibilities later), then, in differential time dt the differential mass increase is, with c some constant,

$$dm = 4\pi r^2 c \, dt;$$

that is,

$$\frac{dm}{dt} = 4\pi r^2 c.$$

Now, the mass of the droplet is, with ρ as the density of water,

$$m = \frac{4}{3}\pi r^3 \rho,$$

---

[1] This is the start of a famous problem in undergraduate physics that has been traced back to the British astronomer James Challis (1803–1882). He posed it as a problem on the 1853 Second Smith's Prize Examination in mathematical physics at Cambridge University.

and so

$$\frac{dm}{dt} = 4\pi r^2 \frac{dr}{dt}\rho,$$

or using our earlier result for $\frac{dm}{dt}$ on the left-hand side, we have

$$\frac{dr}{dt} = \frac{c}{\rho}.$$

Whatever the constants ρ and c are, $\frac{c}{\rho}$ is a constant, and so, with k that constant,

$$\frac{dr}{dt} = k.$$

Thus, integrating,

$$r(t) = kt + C,$$

where C is some constant. Since $r = a$ at $t = 0$, we have $C = a$, and so

$$r(t) = a + kt.$$

Now, Newton's second law of motion says that "force is the rate of change of momentum" and so, with F as the gravitational force on the droplet, equal to mg, where g is the acceleration of gravity, we have

$$F = mg = \frac{d}{dt}(mv) = m\frac{dv}{dt} + v\frac{dm}{dt}.$$

So,

$$m\frac{dv}{dt} = mg - v\frac{dm}{dt},$$

or

$$\frac{dv}{dt} = g - \frac{v}{m}\frac{dm}{dt}.$$

Since

$$\frac{dm}{dt} = 4\pi r^2 \frac{dr}{dt}\rho,$$

we have

$$\frac{dv}{dt} = g - \frac{v}{\frac{4}{3}\pi r^3 \rho}4\pi r^2 \frac{dr}{dt}\rho = g - \frac{3v}{r}k,$$

or

$$\frac{dv}{dt} = g - \frac{3kv}{a+kt}.$$

You can verify, by direct substitution, that this differential equation has the solution

$$v(t) = \frac{g}{4k}\left[a + kt - \frac{a^4}{(a+kt)^3}\right].$$

Notice that this solution says $v(0) = 0$, which agrees with the statement that the droplet starts its fall *from rest*.[2]

This is a remarkable result. Suppose, for example, that the drop begins its existence as a microscopic particle of dust, around which water vapor begins to condense. Then, $a = 0$, and $v(t)$ reduces to

$$v(t) = \frac{gt}{4}.$$

All dependence on the constant k (and so on the constants c and ρ) vanishes. But not only that, the *acceleration* experienced by the falling,

---

[2] At this point you are surely wondering just *where* did the curious $v(t)$ in the box come from? I assure you, I did *not* guess it! The details of its *derivation* would be a distraction here, but at the end of this discussion, as part of the first challenge question, I'll walk you through the steps, steps that every mathematical physicist should have in his or her bag of tricks.

ever-more-massive drop is

$$a(t) = \frac{dv}{dt} = \frac{1}{4}g,$$

which is (significantly) *less* than the acceleration of gravity! This, I feel safe in saying, is a most nonintuitive result.[3]

Of course, we *have* overlooked some practical details, like air drag. Such a detail could significantly complicate the analysis,[4] but the one assumption we've made in particular that might be of at least as much concern is the mass accretion rule. Why should water vapor *uniformly* deposit on the entire surface of the drop? Shouldn't, instead, the front end of the drop, the half of the drop that "runs into the foggy water mist," accumulate more mass than does the rear half of the drop as it falls? And the faster it falls through the fog, the faster the drop should gain mass? If so, the mass accretion rule should be changed to the more realistic

$dm = k$(cross-sectional area of the drop)(speed of the drop)dt,

where k is some constant. That is,

$$\frac{dm}{dt} = \pi r^2 vk,$$

where the water "swept out" by the falling drop is imagined to instantly and continuously distribute itself around the entire drop. Thus, the radius r of the drop increases with time, but certainly in a way different from our first analysis. To analyze this new mass accretion rule is (no surprise) more involved than before, but if we are willing to make the

---

[3]Well, you say, *why* nonintuitive? After all, nearly everybody would *immediately* say that a basketball would fall *very* slowly in a pool of honey, with an acceleration *far* less than g. Yes, of course, I reply, but I would be willing to bet those same people would be surprised at our result for a raindrop falling through a fog. After all, honey is thick and sticky, while a fog is, well, just *wispy*.

[4]But not, surprisingly, to the point where things become impossible. See H. Hossein Partovi and Duane R. Aston, "The Generalized Raindrop Problem," *American Journal of Physics*, October 1989, pp. 912–920.

"drop starts as a microscopic drop" assumption as soon as possible, then the analysis is still not that hard. Here's how that approach works.

With the symbols as before, we have the mass of the drop as

$$m = \frac{4}{3}\pi r^3 \rho,$$

and so

$$\frac{dm}{dt} = 4\pi r^2 \frac{dr}{dt}\rho,$$

or as

$$4\pi r^2 = \frac{3m}{\rho r},$$

then

$$\frac{dm}{dt} = \frac{3m}{\rho r}\frac{dr}{dt}\rho,$$

or

$$\frac{dm}{dt} = \frac{3m}{r}\frac{dr}{dt}. \tag{A}$$

From the new mass accretion rule, we have

$$\frac{dm}{dt} = \pi r^2 vk, \tag{B}$$

and, from Newton's second law of motion, we have

$$mg = m\frac{dv}{dt} + v\frac{dm}{dt}. \tag{C}$$

From (A) and (B) we have

$$v = \frac{1}{\pi r^2 k}\frac{dm}{dt} = \frac{1}{\pi r^2 k}\frac{3m}{r}\frac{dr}{dt} = \frac{3m}{k\pi r^3}\frac{dr}{dt} = \frac{3m}{k\frac{3m}{4\rho}}\frac{dr}{dt},$$

or

$$v = \frac{4\rho}{k}\frac{dr}{dt}. \tag{D}$$

Thus,

$$\frac{dv}{dt} = \frac{4\rho}{k}\frac{d^2r}{dt^2}. \tag{E}$$

Putting (A), (D), and (E) into (C), we get

$$mg = \left(\frac{4\rho}{k}\frac{dr}{dt}\right)\left(\frac{3m}{r}\frac{dr}{dt}\right) + m\left(\frac{4\rho}{k}\frac{d^2r}{dt^2}\right),$$

or canceling the m's and multiplying through by kr and dividing by ρ,

$$\frac{gk}{\rho}r = 12\left(\frac{dr}{dt}\right)^2 + 4r\frac{d^2r}{dt^2}.$$

Now, here's where we make the *assumption* of an initially microscopic origin for the drop. Let's *assume* $r(t) = ct^n$, where c is some constant (that is, the drop's radius grows as some power of t). This says, of course, that $r(0) = 0$. How do we know $r(t) = ct^n$? Well, we really don't, but if we make that *assumption*, all our equations *will* remain consistent, and that is certainly encouraging! Given that we do this, the next obvious question is, what's the value of n? To answer that, we substitute this assumed $r(t)$ into our last equation to get

$$\frac{gk}{\rho}ct^n = 12\left(cnt^{n-1}\right)^2 + 4ct^n\left[cn(n-1)t^{n-2}\right],$$

or

$$\frac{gk}{\rho}ct^n = 12c^2n^2t^{2n-2} + 4c^2n(n-1)t^{2n-2}.$$

For the time behavior to be the same on both sides of the equality it is clear that $n = 2n - 2$, or $n = 2$. That is, $r(t) = ct^2$; the drop's radius increases with the *square* of time. Also,

$$\frac{gk}{\rho}c = 12c^2n^2 + 4c^2n(n-1),$$

or, with $n = 2$ and dividing through by c,

$$\frac{gk}{\rho} = 48c + 8c = 56c,$$

or

$$c = \frac{gk}{56\rho}.$$

That is,

$$r(t) = \frac{gk}{56\rho}t^2.$$

So, using (E), the drop's acceleration is

$$\frac{dv}{dt} = \frac{4\rho}{k}2\frac{gk}{56\rho} = \frac{8}{56}g = \frac{1}{7}g.$$

With the new mass accretion rule, the drop falls even *more slowly* than it did before!

Okay, here are four challenge questions for you to try your hand at.

(a) You'll recall that, in note 2, I said I would show you how to solve the differential equation

$$\frac{dv}{dt} = g - \frac{3kv}{a+kt}$$

for the $v(t)$ in the boxed equation. Here's how to do that.

(1) Write $\frac{dv}{dt} + \frac{3kv}{a+kt} = g$, and then define $P(t) = \int_0^t \frac{3k}{a+kx}dx$, which means

$$\frac{dP}{dt} = \frac{3kv}{a+kt};$$

(2) Form $e^{P(t)} = e^{\int_0^t \frac{3k}{a+kx}dx}$;

(3) Multiply through the differential equation by $e^{P(t)}$ to get

$$\frac{dv}{dt}e^{P(t)} + \frac{dP}{dt}e^{P(t)}v = ge^{P(t)};$$

(4) Notice that the left-hand side of this last equation is $\frac{d}{dt}\left(ve^{P(t)}\right)$, and so

$$\frac{d}{dt}\left(ve^{P(t)}\right) = ge^{P(t)};$$

(5) Integrate both sides of this last equation to get
$ve^{P(t)} - C = \int_0^t ge^{P(y)}dy$ with C some constant. Thus, in general,

$$v(t) = Ce^{-P(t)} + ge^{-P(t)}\int_0^t e^{P(y)}dy;$$

(6) Since $v(0) = 0$, and $P(0) = 0$, we have $0 = Ce^0 + ge^0 \int_0^0 e^{P(y)}dy = C$;

(7) Thus, for $v(0) = 0$, the solution is $v(t) = ge^{-\int_0^t \frac{3k}{a+kx}dx} \int_0^t e^{\int_0^y \frac{3k}{a+kx}dx} dy$.

Your challenge is to perform the indicated integrations in (7) and show that the result is, indeed, the $v(t)$ I gave you.[5]

(b) When Professor Challis created the falling raindrop problem (see note 1) for the 1853 Smith's Prize Examination, he wasn't totally sadistic. He actually gave the students the answer, and "simply" asked them to show it. Specifically, he wrote: "A spherical rain-drop descending under gravity, receives continually by precipitation of vapour an accession of mass proportional to its surface; *a* being its

---

[5]This method, of forming $P(t)$ and then multiplying through the differential equation by $e^{P(t)}$, is called the *method of integrating factors*. It is *enormously* useful in solving what mathematicians call *first-order linear differential equations*, and you can read more about the technique in any good book on differential equations. It is a routine tool for all mathematical physicists (indeed, this is the method used by Herreman and Pottel to *exactly* solve the problem of sliding on a hemisphere with friction, discussed in Problem 14).

radius when it begins to descend, and r its radius after the interval t, show that its velocity is given by the equation

$$v = \frac{gt}{4}\left(1 + \frac{a}{r} + \frac{a^2}{r^2} + \frac{a^3}{r^3}\right),$$

the resistance of air being left out of account." This is a particularly nice form, as it shows that if $r = a$ (that is, if the radius of the drop is *fixed* at the initial radius for all t), then $v = gt$, and so, for a drop that does not gain mass as it falls, the acceleration is a full g, just as you'd expect. Your challenge is to show that the $v(t)$ in the boxed equation (and that you derived in the first challenge) is equivalent to Professor Challis's power series solution.

(c) For this challenge question, you are to determine what happens under yet another mass accretion rule. Specifically, imagine an initially microscopically small drop at the top of a fog bank that is of ever-increasing vapor density as one descends through it. That is, as the drop falls toward the ground it encounters a *linearly* increasing fog density. If we assume that the drop picks up mass both in proportion to its cross-sectional area and the distance it has fallen, and we measure the distance fallen by x (and so $x(0) = 0$), then the mass accretion rule is, with k some constant,

$$\frac{dm}{dt} = k\pi r^2 x,$$

where $r(0) = 0$. The other equations that define the physics of this falling drop are (with ρ the density of water):

$$m = \frac{4}{3}\pi r^3 \rho, \quad \frac{dm}{dt} = \frac{3m}{r}\frac{dr}{dt}, \quad v = \frac{dx}{dt}, \quad mg = m\frac{dv}{dt} + v\frac{dm}{dt}.$$

Find the acceleration of the drop. *Hint* 1: Again, it's a constant, and it's *less* than $\frac{1}{7}g$. *Hint* 2: To keep from being driven crazy with messy equations, I *strongly* suggest you use Newton's dot notation for time derivatives (see Appendix 3), where $\dot{r}\frac{dr}{dt}, \ddot{r} = \frac{d^2r}{dt^2}$, and so on. That is, use

$$\dot{m} = \frac{3m}{r}\dot{r}, \quad \dot{m} = k\pi r^2 x, \quad v = \dot{x}, \quad mg = v\dot{m} + m\dot{v}.$$

(d) For the final challenge in this discussion, here's a mass accretion rule that's even more complicated than the one in (c). As before, the initially microscopically small drop starts at rest at the top of a fog bank that is of linearly increasing vapor density, but now, as it falls the drop picks up mass in proportion not only to its cross-sectional area and the distance it has fallen but also in direct proportion to its speed v. That is, replace the mass accretion rule of (c) with the rule $\frac{dm}{dt} = k\pi r^2 xv$. Find, for this new rule, the acceleration of the drop. *Hint* 1: It's even less than the result in (c). *Hint* 2: Based on the earlier results, *assume* that $r(t) = ct^n$, as well as *assume* that the acceleration of the drop is $\dot{v} = ag$, where a is some constant. Your problem, then, is to show that under these assumptions there is a unique value (find it!) for a.

PROBLEM 19

Beyond the Quadratic: A Cubic Equation and Discontinuous Behavior in a Physical System

If the writers of the *Boston Globe* letter were perplexed by the usefulness of quadratic equations, what (do you wonder) would they have to say about *cubic* equations? Nothing good, I suspect. As mathematical physicists, however, we have to be "armed and ready" for such things because cubic equations do occur (routinely) in physics.[1] Figure P19.1 shows such a situation, one with a surprising conclusion hidden away in

---

[1] In the interest of full disclosure, I should tell you that the famous mathematical physicist Richard Feynman once made what I think an odd comment about cubic equations, a comment that the authors of the *Boston Globe* letter might argue supports their position. Writing in his equally famous book *What Do You Care What Other People Think?* W. W. Norton, 2001, p. 95, he included this remark about the educational attitudes he observed while on a visit in 1980 to Greece: "They were very upset when I said the development of the greatest importance to mathematics in Europe was the discovery [in the 16th century] by Tartaglia that you can solve a cubic equation: *although it is of little use in itself* [my emphasis]." Feynman was known for saying provocative (even outrageous) things simply for the sake of starting a debate, and I think this is an example of that. What upset the Greeks was Feynman's apparent dismissal of the supreme importance of the *ancient Greek* mathematicians in favor of a "modern" Italian!

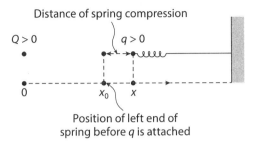

Figure P19.1. Compression of a spring via electrical repulsion.

it. That figure shows a one-dimensional coordinate system, with $x = 0$ at the far left. Located there is a positive electric charge with fixed value Q, firmly fastened to $x = 0$. (If it helps, think of Q being "glued" to $x = 0$.) At the far right, we have a relaxed spring whose unstretched or compressed length puts its left end at $x = x_0 > 0$. The right end of the spring is attached to an unmovable wall. Now, suppose we attach an electric charge q of arbitrary sign and value to the left end of the spring; that is, q can be either positive or negative, and of any value. What happens as we try different values and signs of q?

As everybody remembers from high school physics, electrical charges of opposite sign attract, while charges of the same sign repel. In Figure P19.1 we see the case for $q > 0$, which shows that the left end of the spring has moved *away* from Q. Q can't move—remember, it's glued to $x = 0$—but q can move under the force of repulsion. As q moves to the right, the spring compresses. That compression results in the spring offering increasing resistance to further compression. Of course, as q moves away from Q the electrical repulsion force decreases. It would seem, then, that there should be some value of x (the location of q) at which the electrical repulsion force and the mechanical compression force balance, and so q will eventually come to rest at that particular x. This is correct for the $q > 0$ case, but to understand *quantitatively* what happens, and particularly so for the $q < 0$ case, we need to do some mathematical physics.

To start, let's continue with $q > 0$ and a compressed spring. Since the spring has shortened in length by $x - x_0$, the mechanical force on q from the spring pushing q back toward Q is, by *Hooke's law*,[2] given

[2]Named after the English scientist Robert Hooke (1635–1703).

by

$$c_1(x - x_0),$$

where c_1 is some positive constant (called the *spring constant*). The electrical repulsion force on q, due to Q, is given by *Coulomb's law*,[3] given by

$$c_2 \frac{Qq}{x^2},$$

where c_2 is some positive constant that depends on the electromagnetic details of the space in which Q and q reside.

As discussed earlier, a plausible answer to the question, "what happens?" in the $q > 0$ case is that q retreats from Q until the repulsion force due to Q is just balanced by the mechanical resistive force of the compressed spring. At that point, q stops moving, and we say that Q, q, and the spring have arrived at a *state of equilibrium*. The value of x for that state is the solution to

$$c_1(x - x_0) = c_2 \frac{Qq}{x^2}.$$

That is, if we write q as a function of x,

$$q = \frac{c_1}{c_2 Q} x^2 (x - x_0). \tag{A}$$

The numerical value of x_0 simply sets the physical scale of Figure P19.1, and that is something we can do for our convenience. So, let's pick the convenient value $x_0 = 1$ (if we are working in the MKS system, this means x_0 is 1 meter). Also, let's pick Q to be whatever value is required to make

$$\frac{c_1}{c_2 Q} = 1.$$

Again, if we are in the MKS system, then Q and q are both measured in units of coulombs. With these choices, (1) becomes

$$q = x^2 (x - 1), \tag{B}$$

---

[3] Named after the French scientist Charles Augustin de Coulomb (1736–1806).

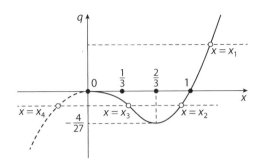

Figure P19.2. Spring charge (q) versus distance (x).

a cubic equation that contains all the physics of the arrangement of Figure P19.1.

There are, of course, analytical ways[4] to solve (2), to find x for a given q, but a mathematical physicist stranded on a remote island, with no calculator or math books, could still find out everything he or she needs to know using *nothing* but a stick to scribble with on a sandy beach. With that stick and sand to write in, these observations follow:

(a) $q = 0$ at $x = 0$ and $x = 1$;

(b) $q \leq 0$ for $0 \leq x \leq 1$;

(c) $q > 0$ for $x > 1$;

(d) since $\frac{dq}{dx} = 3x^2 - 2x$, then

(e) $\frac{dq}{dx} = 0$ at $x = 0$, and $x = \frac{2}{3}$ (where $q = -\frac{4}{27}$);

(f) $\frac{dq}{dx} = 1$ at $x = 1$;

(g) since $\frac{d^2q}{dx^2} = 6x - 2$, then

(h) $\frac{d^2q}{dx^2} < 0$ at $x = 0$, and $\frac{d^2q}{dx^2} > 0$ at $x = \frac{2}{3}$.

From these simple observations, Figure P19.2 immediately follows, where (e) and (h) tell us that a plot of q versus x has a local *maximum* at $x = 0$ and a local *minimum* at $= \frac{2}{3}$.

There is a tremendous amount of information packed into Figure P19.2, including that surprise I hinted at earlier. I'll start by directing

[4]See Appendix 3.

your attention to the upper horizontal dashed line through an arbitrary $q > 0$ value. That line intersects the cubic plot at $x = x_1$, and this shows that for *any* $q > 0$ there is always exactly *one* positive solution to (2), and that the larger is q the greater is x_1. This makes physical sense: as we increase q from one positive value to a larger positive value, the force of Q's repulsion increases, and so the spring compresses a bit more as q moves a bit farther away from Q.

For $q < 0$, however, we suddenly find a distinctly different behavior. If $-\frac{4}{27} < q < 0$, we see that there are *two* positive solutions to (2), at $x = x_2$ and $x = x_3$, as well as a third, *negative* solution at $x = x_4$. While $x = x_4$ does solve (2), we immediately reject it on physical grounds, as it declares q to be on a spring extended *beyond* Q. That still leaves us with two solutions, however, and what could that *mean*, physically? And perhaps even more puzzling is that for $q < -\frac{4}{27}$ there are *no* solutions (a horizontal line through $q < -\frac{4}{27}$ does not intersect the cubic plot for any positive x)! What could *that* mean? Surely, after all, if we attach a charge of, say, $q = -\frac{5}{27}$ to the spring, *something* will happen—but what? Figure P19.2 doesn't seem to provide an answer.

We can address all these puzzles by adding the idea of *stability* to the concept of an equilibrium state. An equilibrium state is *stable* if a small perturbation to it (changing x to $x - \epsilon$, for example, where ϵ is *very* small but not zero) produces effects that counter the perturbation. The equilibrium state is *unstable* if such a perturbation leads to effects that enhance the perturbation. (Recall the discussion in Problem 14 of sliding on a hemisphere of frictionless ice.) With this idea, we can show that $x = x_1$ for $q > 0$ is a stable state, as is $x = x_2$ for $q < 0$, but $x = x_3$ for $q < 0$ is an unstable state. So, while the $x = x_3$ state satisfies the mathematics of Figure P19.1, it does not satisfy the physics. Here's how to show that these claims are true.

The first state, $x = x_1$ for $q > 0$, is easy to analyze without using any math at all. As Figure P19.1 shows, $x_1 > 1$ (the spring is compressed), and so, taking $\epsilon > 0$, moving q to $x = x_1 - \epsilon$ slightly *decompresses* the spring. That is, the spring exerts a *smaller* force on q after the perturbation. At the same time, the electric repulsion force *increases*. Both effects work together to move q back toward $x = x_1$, and so we conclude that this state is a stable state. If we take $\epsilon < 0$, we'll reach the same conclusion.

The situations for the two states with $q < 0$ ($x = x_2$ and $x = x_3$) are a bit more subtle, and now we will need to do a bit of math. First, consider $x = x_2$, where $\frac{2}{3} < x_2 < 1$ (the spring is now stretched), and then we alter q's location to $x = x_2 - \epsilon$, where $\varepsilon > 0$. Since the spring was already stretched at $x = x_2$, the perturbation has increased the spring's length by ϵ and so the mechanical spring force (attempting to pull q back to $x = x_2$) has *increased* by $c_1\epsilon$. The electrical *attraction* force (remember, $q < 0$ while $Q > 0$) before the perturbation is $c_2 \frac{Qq}{x_2^2}$. After the perturbation this force is $c_2 \frac{Qq}{(x_2-\epsilon)^2}$. Recalling that $q < 0$, $c_2 \frac{Qq}{(x_2-\epsilon)^2}$ is *more negative* than $c_2 \frac{Qq}{x_2^2}$, and so the *positive* change in the *attraction* force that is attempting to move q toward Q (and so even farther away from $x = x_2$) is

$$
c_2 \frac{Qq}{x_2^2} - c_2 \frac{Qq}{(x_2-\epsilon)^2} = c_2 Qq \left[\frac{1}{x_2^2} - \frac{1}{(x_2-\epsilon)^2} \right] = c_2 Qq \left[\frac{(x_2-\epsilon)^2 - x_2^2}{x_2^2(x_2-\epsilon)^2} \right]
$$

$$
= c_2 Qq \left[\frac{x_2^2 - 2\epsilon x_2 + \epsilon^2 - x_2^2}{x_2^2(x_2^2 - 2\epsilon x_2 + \epsilon^2)} \right] \approx -c_2 Qq \left[\frac{2\epsilon}{x_2^3} \right]
$$

$$
= -c_1 \left(\frac{c_2 Q}{c_1} \right) q \frac{2\epsilon}{x_2^3} = -c_1 q \frac{2\epsilon}{x_2^3},
$$

an expression which becomes ever more correct as $\epsilon \to 0$. Note, carefully, that this last expression *is* positive because $q < 0$.

Since

$$
q = x_2^2(x_2 - 1),
$$

the positive change in the attractive force on q is

$$
-c_1 x_2^2(x_2 - 1) \frac{2\epsilon}{x_2^3} = c_1 x_2^2(1 - x_2) \frac{2\epsilon}{x_2^3} = 2c_1 \epsilon \frac{1 - x_2}{x_2}, \quad \frac{2}{3} < x_2 < 1.
$$

The question of interest now is clear: which is greater, $c_1 \varepsilon$ (which would mean the equilibrium state $x = x_2$ is stable) or $2c_1 \varepsilon \frac{1-x_2}{x_2}$ (which would mean the equilibrium state is unstable)? The answer is that the $x = x_2$

is stable because, defining Δ as the increase in the spring force minus the increase in the electric force, we have

$$\Delta = c_1\varepsilon - 2c_1\varepsilon\frac{1-x_2}{x_2} = c_1\varepsilon\left[1 - 2\frac{1-x_2}{x_2}\right] = c_1\varepsilon\left[\frac{x_2 - 2 + 2x_2}{x_2}\right]$$

$$= c_1\varepsilon\left[\frac{3x_2 - 2}{x_2}\right] > 0$$

for all x_2 in the interval $\frac{2}{3} < x_2 < 1$.

For the $x = x_3$ state we have $0 < x_3 < \frac{2}{3}$, and so now we find $\Delta < 0$ for all x_3 in that interval, and thus the $x = x_3$ state is unstable. We now have the answer to the puzzle of what *two* equilibrium states for the $q < 0$ case mean: only *one* of them is physically meaningful (the stable $x = x_2$ case). But what of the puzzle of Figure P19.2, seemingly not allowing q to be more negative than $-\frac{4}{27}$? What prevents us from making q *anything* we wish? The concept of state stability provides the answer too.

Imagine that we have $q < 0$, but it's not quite yet *at* $q = -\frac{4}{27}$. As we continue to make q more negative, we slide down the cubic in Figure P19.2 toward the local minimum at $x = \frac{2}{3}$. As we do, the electrical attraction force due to Q continually increases but, as we showed in our stability analysis, not enough to dominate the increasing mechanical restraining force of the continually stretching spring. Then, when q reaches $-\frac{4}{27}$ these two forces *just* balance. And then, if we make q *infinitesimally* more negative than $-\frac{4}{27}$, the attractive electrical force at last exceeds the restraining spring force, and because now q is in the *unstable* state portion of the cubic, q *continues* to move toward Q. Indeed, q almost instantly *jumps* from $x = \frac{2}{3}$ to $x = 0$ (where, theoretically, the electrical attractive force becomes infinitely large).

When we had $-\frac{4}{27} < q < \infty$, we observed that a *continuous* change in q gave a *continuous* change in x. When we attempted to move q beyond $-\frac{4}{27}$ to more negative values, however, x changed in a *discontinuous* way from $x = \frac{2}{3}$ to $x = 0$. This is the "surprise" I mentioned earlier, one that is inherent in what *appears* to be simple physics—but actually isn't so simple.[5] Similar sudden changes (called *phase transitions*) in physical systems occur in many everyday situations: for example, a continuous decrease in the temperature of a glass of water seems to result in simply an ever-colder glass of *water* until, at a critical temperature, the *liquid water* suddenly becomes *solid ice*. Another commonly observed phase transition occurs when one more snowflake is *the one* that causes a previously intact limb to suddenly snap off a tree trunk.

Your challenge question for this problem is to determine the stability of the $x = x_1$, $x = x_2$, and $x = x_3$ states in the charge/spring system using a different (and more general) approach than the one I used. This new approach is to examine the *potential energy* of the system in its equilibrium states and to argue that for a *stable* state the potential energy is a local *minimum*, while for an *unstable* state the potential energy will be a local *maximum*. The fundamental idea is that a physical system, left to itself, will always settle into the state with the smallest possible potential energy. (Physics has a number of so-called extrema principles, and this is one of them. Such principles have a metaphysical/philosophical flavor to them, but physicists generally accept them because they often lead to correct solutions.) Before tackling the charge/spring system, let me first show you how this works in a simpler situation.

Imagine a system consisting of a solid rectangular block of matter with cross-sectional area A and height l, with density ρ_s, positioned so its bottom surface is just on the surface of a pool of liquid of infinite extent, as shown in Figure P19.3. Let's call this state of the solid-block/liquid-pool system the *zero-energy state*; that is, it is our reference state, against which we'll compare all other states of the block/pool system.

If the fluid has density $\rho_f > \rho_s$, we know what will happen if the zero-energy state is allowed to adjust on its own. The block will partially sink into the fluid until it reaches the equilibrium state called *floating*

---

[5]An interesting modification of this problem, suitable for experimentation at home, is described in the paper by M. Partensky and P. D. Partensky, "Can a Spring Beat the Charges?" *The Physics Teacher*, November 2004, pp. 472–476. There, Q and q are replaced by the north and/or south poles of high-strength magnets. Unlike the electric charge Q, a magnet *can* be glued to the origin!

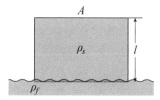

Figure P19.3. The zero-energy state for a liquid pool/solid block system.

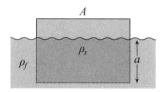

Figure P19.4. The floating state.

(think of a wooden block in a pool of water). Let's suppose, in fact, that the block sinks distance a, as shown in Figure P19.4. Thus, the potential energy of the *block alone* will *decrease*. But what happens to the displaced fluid? Every molecule of fluid in the pool can be imagined as remaining right where it was before the block partially sank, except for the molecules that were displaced, which can be imagined as spread uniformly over the surface of the pool. That is, the mass of the displaced fluid is elevated, and so the potential energy of the fluid alone will *increase*.

Let's now work out these two changes in potential energy and so find the new energy (relative to the zero-energy state) of the floating state. The block is easy. The mass of the block is $Al\rho_s$, and so after it sinks distance a into the pool, the potential energy *change* for the block is $-Al\rho_s ga$, where g is the acceleration of gravity. The potential energy calculation for the displaced fluid is just a bit more involved (but not by much).

The sinking block displaces a volume of fluid in the amount of Aa, and so the mass of the displaced fluid is $Aa\rho_f$. The center of mass of the displaced fluid is originally at depth $\frac{1}{2}a$ (at the midpoint of the displaced rectangular volume of fluid), and so when the displaced fluid appears as an *infinitesimally thin* film over the surface of the pool (remember, the pool is of infinite extent), all the fluid mass is at the same elevation and so, too, is the center of mass. That is, the center of mass of the displaced fluid has risen by $\frac{1}{2}a$, and so the potential energy change for the fluid is an *increase* of

$$Aa\rho_f g \frac{1}{2}a = \frac{1}{2}Ag\rho_f a^2.$$

Thus, the total potential energy U of the pool/block system for the floating state is

$$U = \frac{1}{2}Ag\rho_f a^2 - Al\rho_s ga.$$

Notice that $U = 0$ when $a = 0$, just as it should, since $a = 0$ is how we *defined* the zero-energy state. We can easily solve this energy equation for a because, as you'll also notice, it's a quadratic equation (and so we once again thrust a sword into the breast of the *Boston Globe* letter argument):

$$a = \frac{l\rho_s}{\rho_f} \pm \frac{1}{\rho_f}\sqrt{l^2\rho_s^2 + U\frac{2\rho_f}{Ag}}.$$

Now, I think we'd all agree that the value of a must be real, as there is no physical significance to the block sinking into the liquid pool to a complex depth! This means the quantity under the square-root sign must be nonnegative. That is, it must be true that

$$U \geq -\frac{l^2\rho_s^2 Ag}{2\rho_f}.$$

The absolutely *minimum* possible potential energy is, therefore,

$$U = -\frac{l^2\rho_s^2 Ag}{2\rho_f},$$

for which our result for a reduces to

$$a = \frac{l\rho_s}{\rho_f}.$$

This value for a, by our minimum potential energy criterion for a stable floating state, becomes, if we multiply through by $Ag\rho_f$, $Aa\rho_f g = Al\rho_s g$. But the left-hand side is the *weight* of the displaced fluid, and the right-hand side is the *weight* of the block. That is, our minimum potential energy principle for a stable state has led us directly to Archimedes' famous result.

To summarize, if $U(x)$ is the potential energy of a physical system, then for an equilibrium state of that system at $x = x^*$, we'll find that

$$\frac{dU}{dx}\big|_{x=x^*} = 0 \text{ and } \frac{d^2U}{dx^2}\big|_{x=x^*} > 0 \text{ if } x = x^* \text{ is stable,}$$

while

$$\frac{dU}{dx}\big|_{x=x^*} = 0 \text{ and } \frac{d^2U}{dx^2}\big|_{x=x^*} < 0 \text{ if } x = x^* \text{ is unstable.}$$

Now, back to the charge/spring system.

Since the spring energy is given by $\frac{1}{2}c_1(x - x_0)^2$, and the electric field energy[6] is given by $c_2\frac{Qq}{x}$, then for our problem (setting $x_0 = 1$ and $c_2 Q = c_1$) we have

$$U(x) = c_1 \left[\frac{q}{x} + \frac{1}{2}(x - 1)^2 \right].$$

---

[6]The spring energy is found by computing the work (energy) integral $|\int F dx|$, where $F = c_1(x - x_0)$. The electric field energy is found by computing the work (energy) integral $-\int_{\infty}^{x} F dr$, where $F = c_2\frac{Qq}{r^2}$. By convention, we imagine q being transported in the electric field of Q from plus infinity to its position at the left end of the spring. The minus sign gives a positive energy when $q > 0$ is transported through the *repulsive* field of $Q > 0$.

Therefore,

$$\frac{dU}{dx} = c_1 \left[-\frac{q}{x^2} + (x-1) \right],$$

and so

$$\frac{dU}{dx} = 0$$

when

$$-\frac{q}{x^2} + (x-1) = 0.$$

That is, $\frac{dU}{dx} = 0$ at those values of x such that $q = x^2(x-1)$. In other words, $\frac{dU}{dx} = 0$ at the *same values* of x that are the equilibrium states of the system. Now, finish the analysis by computing $\frac{d^2U}{dx^2}$ at the equilibrium states $x = x_1$, $x = x_2$, and $x = x_3$.

Springs Can Be Lethal

A discussion of springs might seem to be more in the realm of engineering than physics, but springs can help mathematical physicists understand the spectacular nature of what are called *kinetic energy explosions*. A kinetic energy explosion occurs when a very high-speed mass (like an asteroid) *suddenly smashes* to a halt. If we *rolled* an asteroid into the ocean, for example, all that would happen would be a big splash and a lot of water would be displaced. However, if that same asteroid *hit* the ocean at 20,000 m/s, we'd get a massively violent detonation that would easily dwarf, many times over, an explosion of the largest hydrogen bomb. The difference is that when *rolled* into the ocean, water simply flows away from and around the rock. But a rock traveling at 20 kilometers per second hits *so fast* that water simply can't move quickly enough to get out of the way—instead, the water's molecular bonds are literally compressed (like little springs), storing the kinetic energy of the asteroid as it is brought to a stop. Then those springs rebound, unleashing all of the asteroid's KE in a sub-microsecond instant. (The release of energy in such a short time is a characteristic of *nuclear* explosions: *chemical* explosions are a thousand times or more *slower*.) And it can be a *lot* of kinetic energy. The asteroid that hit Earth 65 million years ago, for example, the one that killed off the dinosaurs, is estimated to have been 10 kilometers in diameter (imagine standing next to a spherical rock whose top is 10,000 m up in the sky!) and to have released an energy equivalent to 100 million megatons of TNT. By comparison, the largest nuclear test explosion, ever, was from a Russian device set off in 1955, estimated to have produced a "mere" 50 megaton blast. Even a much smaller body than the "dinosaur killer" can produce an enormous explosion: Meteor Crater in Arizona was created 50,000 years ago by a mass only 50 meters in diameter, but it produced a kinetic energy blast of about 29 megatons. All this from 'compressed springs.' For much more on the physics of asteroid impacts, see the fascinating book by David Shonting and Cathy Ezrailson, *Chicxulub: The Impact and Tsunami (The Story of the Largest Known Asteroid to Hit the Earth*, Springer, 2017).

PROBLEM 20
Another Cubic Equation: This One Inspired by Jules Verne

In the previous problem you saw a cubic equation appearing in an electrical physics situation, and here I'll show you a cubic appearing in a gravitational setting. The two physical phenomena are simultaneously strikingly different (electrical charges can both attract and repel, while masses *always* attract), as well as similar (the electrical and gravitational forces both obey inverse-square laws). Because of the central theme of this book I'll concentrate here on physics, but don't think that mathematicians have grown weary of cubics just because they were "solved" centuries ago. Cubic equations, in fact, continue to be at the heart of some major questions that mathematicians are wrestling with as you read this.[1]

But, back to physics. As the author of one recent paper observed, a common problem in freshman physics textbooks is the calculation of the location in space between two *point* masses where the net gravitational force on a so-called test mass is zero.[2] This question generally appears with no motivation, other than as an exercise in the algebra of the inverse-square law of gravity. Professor Lima (note 2) mentions that, occasionally, students suggest that such a thing (the opposing gravitational forces of the Earth and the Moon) can "explain" why astronauts in orbit are weightless, but he quickly puts that faulty idea to rest. (The correct explanation for why we occasionally see astronauts on the evening news floating about in the International Space Station is that the inward gravitational force due to Earth is just equal to the outward "centrifugal force" on the astronaut as he/she travels along a circular orbit.[3])

Long predating actual space travel, however, is a *literary* motivation for calculating the location of the place where gravity vanishes between two masses, a motivation that Professor Lima doesn't mention: the wonderfully fantastic 1865 novel *From the Earth to the Moon*, by the French pioneer in science fiction, Jules Verne (1828–1905). There he introduced his readers to the then-startling concept of a weightless space traveler, as a result of the cancellation of opposing gravitational forces. As Verne imagined things, after the tale's fabulous 900-foot-long cannon detonated its charge of 200 tons of guncotton to blast a manned capsule into space at an initial speed of 12,000 yards per

[1] See, for example, Barry Cipra, "Fermat Prover Points to Next Challenges," *Science*, March 22, 1996, pp. 1668–1669. There you'll find some discussion on the continuing mysteries of what are called *elliptic curves*, described by cubic equations of the form $y^2 = x^3 + Ax + B$, where A and B are integers. The central question is, given A and B, does such an equation have rational solutions for x and y?

[2] F.M.S. Lima, "Where Else Is Null the Gravitational Field Between Two Massive Spheres?" *European Journal of Physics*, July 2009, pp. 785–792.

[3] My use of the word *centrifugal* will cause many physicists to grind their teeth. They will say that I should instead say that the Earth's gravitational force is the *centripetal* force that explains the acceleration of the astronauts as they literally "fall around the Earth" on a circular orbit.

second (6.8 miles per second) toward the Moon, the capsule would eventually reach "the instant that the attraction of the moon exactly counterpoises that of the earth, that is to say at 47/52 (0.9038) of its passage. At that instant *the projectile will have no weight whatever* [my emphasis]."[4]

Verne strove for verisimilitude in his tales by including a great deal of technical detail, much of it accurate (particularly the mathematics). I don't know the particular details of how Verne arrived at the ratio 47/52, but it isn't difficult to show that the value of that fraction is quite close to the correct value. In Figure P20.1 we see Verne's capsule (with mass μ) at distance p from the Earth's center (with the Moon's center at distance d). Now we make the assumption that μ is in the space *between* the Earth and the Moon (and not inside either one), and so we can use Newton's discovery that, *outside* of a spherically symmetric mass, gravity is the same as it would be if all the mass were concentrated at the center of the sphere. (I'll say more about this later in this discussion.) Since the Earth has 81 times the mass of the Moon,[5] setting the two (oppositely directed) gravitational forces on the test mass μ equal, we have (with M as the Moon's mass) and G as the universal constant of gravitation,[6]

$$G\frac{81M\mu}{p^2} = G\frac{M\mu}{(d-p)^2},$$

or

$$\frac{81}{p^2} = \frac{1}{(d-p)^2},$$

---

[4] From the answer to the first question in Chapter 4 of the novel ("Reply from the observatory of Cambridge"). When Verne writes "of the distance" it is not clear if he means center to center or surface to surface. But, since the Earth's radius is 4,000 miles, and the Moon's is 1,000 miles, while the Moon is about 240,000 miles distant no matter how you measure it, the distinction is actually a matter of only a couple of percent or so.

[5] For how we know this, see my book *Mrs. Perkins's Electric Quilt*, Princeton University Press, 2009, pp. 175–179.

[6] For more on G, see *Mrs. Perkins's Electric Quilt* (note 5), pp. 136–140.

Figure P20.1. Between the Earth and the Moon, at zero gravity.

and so

$$\frac{9}{p} = \frac{1}{d-p},$$

which immediately gives

$$\frac{p}{d} = 0.9,$$

a value pretty close to Verne's 0.9038.

Five years later (1870) Verne wrote a sequel (*Round the Moon*) that picks up the story of the capsule after blastoff, which is where *From the Earth to the Moon* ends. In that sequel more is said about weightlessness, but for some reason Verne changed the ratio of $\frac{p}{d}$ to $\frac{47}{60} = 0.7833$. Why, I don't know. In *Round the Moon* Verne incorrectly wrote of the zero gravity point at which the capsule might be forever trapped.[7] Verne missed the fact that the zero gravity point is one of *unstable* equilibrium: the slightest perturbation of the capsule would have increased that perturbation even more by the gravitational attraction of the massive sphere it happened initially to move toward.

Now, just what does all this have to do with cubic equations? you ask. Well, if Verne had explored the gravitational force between two massive spheres in just a little bit more detail, he would have encountered a cubic. To see this, look at Figure P20.2, which shows two spheres, of masses M and m, with radii R and r, respectively, with their centers

---

[7]See item 3 in Chapter 8, "At Seventy-Eight Thousand Five Hundred and Fourteen Leagues," where Verne writes that at the zero-gravity point between Earth and Moon, "it [the capsule] would remain forever suspended in that spot." The men in Verne's capsule could, however, have easily introduced a saving perturbation by simply letting some air out (as they did when ejecting the body of one of their dogs, killed by the shock of the blastoff).

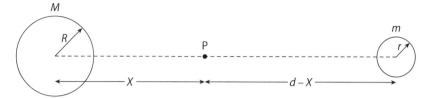

Figure P20.2. Zero gravitational force at P.

distance d apart (with $d > R + r$). The point P on the line joining the two centers is where the gravitational force on a test mass due to M is just balanced by the gravitational force on the test mass due to m. Our question is, where is P?

If we set the center of M at $x = 0$, and P at $x = X$, then.

$$\boxed{\frac{GM\mu}{X^2} = \frac{GM\mu}{(d - X)^2}.}$$

Solving for X, we get

$$X = \frac{d}{1 + \sqrt{\frac{m}{M}}}.$$

This solution for X appears so quickly that it is easy to be fooled into blindly accepting it (with grateful thanks for the no-pain calculation!). Alas, there is a subtle flaw: if the value of X is such that either $X < R$ or $X > d - r$, that is, if X is such that P lies inside either M or m, then *our boxed equation is simply not valid*. The result *is* true for M and m as *point* masses, but for spatially extended spherical masses with nonzero radii, the gravitational forces on a test mass fall off as the inverse square only *outside* the spheres.

To help you understand the physics of this, I'll remind you of Newton's two so-called superb theorems that appear in his monumental *Principia* (1687). I'll be discussing the superb theorems in terms of a single sphere with radius R. The variable x will, in this material, be the distance from the center of that sphere. In the *Principia* Newton showed that a uniformly dense sphere of mass M and radius R acts just like a *point* mass with the mass of the sphere concentrated at the sphere's center *when it interacts with other masses at distance $x \geq R$ from the center of the sphere*. (What happens when $x < R$ we'll take up in just a bit.) An equally stunning result was Newton's discovery that the gravitational force on a point mass inside a hollow spherical shell of matter, no matter where inside the shell the point mass might be located, is *zero*. These two results were developed by Newton in tour de force geometric proofs, but today they are universally proved in freshman physics using freshman calculus.[8]

Now, since Newton's second superb theorem says that the gravitational force on a test mass inside a thin spherical shell of uniform density is zero, then as a test mass descends into the interior of a solid sphere of matter of uniform density (more generally, the density can vary as long as it is spherically symmetric) the inverse-square gravitational force the test mass experiences depends only on the matter still *beneath* it. The matter that is *above* it (the spherical shell it passed through) contributes nothing to the gravitational force on it. Thus, for a sphere with radius R and density ρ, the gravitational force on a test mass μ varies with x as

$$F(x) = \begin{cases} \dfrac{GM\mu}{x^2}, x \geq R \\[2mm] \dfrac{G\frac{4}{3}\pi x^3 \rho \mu}{x^2} = G\dfrac{4}{3}\pi\rho\mu x, x \leq R. \end{cases}$$

Since

$$M = \frac{4}{3}\pi R^3 \rho,$$

then

$$\frac{4}{3}\pi\rho = \frac{M}{R^3},$$

---

[8]See, for example, *Mrs. Perkins's Electric Quilt* (note 5), pp. 140–147.

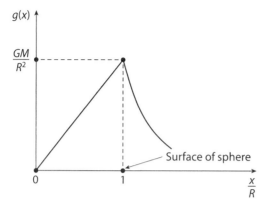

Figure P20.3. Acceleration of gravity inside and outside of a uniform density, solid sphere of radius R and mass M.

and so

$$G \frac{4}{3} \pi \rho \mu x = G \frac{M \mu}{R^3} x.$$

Thus,

$$F(x) = \begin{cases} \frac{GM\mu}{x^2}, & x \geq R \\ \frac{GM\mu}{R^3} x, & x \leq R \end{cases}.$$

Let's write $g(x)$ as the acceleration of gravity at the test mass, distance x from the center of the sphere. Since Newton's famous equation $F = ma$ says

$$F(x) = \mu g(x),$$

then

$$g(x) = \begin{cases} \frac{GM}{x^2}, & x \geq R \\ \frac{GM}{R^3} x, & x \leq R \end{cases}.$$

Figure P20.3 shows how $g(x)$ varies with x (notice, in particular, that $g(x)$ has its maximum value at the surface of the sphere).

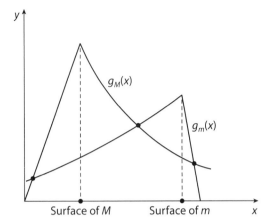

Figure P20.4. There are *three* places where gravity vanishes (between M and m, inside M, and inside m) when the maximum acceleration due to m (on the surface of m) is greater than the acceleration due to M on the surface of m.

Okay, let's now return to the *two* spheres of Figure P20.2 with masses M and m (and radii R and r, respectively), with a center-to-center separation of d. The acceleration of gravity at distance x from the center of mass M, due to M, is

$$g_M(x) = \begin{array}{ll} \frac{GM}{x^2}, & x \geq R \\ \frac{GM}{R^3}x, & x \leq R \end{array},$$

and the acceleration of gravity at distance $d - x$ from the center of mass m, due to m, is

$$g_m(x) = \begin{array}{ll} \frac{Gm}{r^3}(d - x), & d - r \leq x \leq d \\ \frac{Gm}{(d-x)^2}, & x \leq d - r \end{array}.$$

Superimposing plots of these expressions for g_m and g_M quickly shows the possibilities for the zero gravity points (see Figures P20.4 and P20.5).

We know there is always a zero gravity point inside M and wish to compute its location. We set the value of g_M *inside* M equal to the value

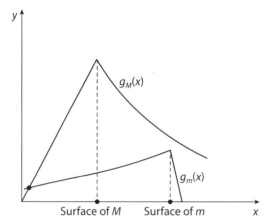

Figure P20.5. There is *one* place where gravity vanishes (inside of M) when the maximum acceleration due to m (on the surface of m) is less than the acceleration due to M on the surface of m.

of g_m *outside m*. Thus,

$$\frac{GM}{R^3}x = \frac{Gm}{(d-x)^2},$$

which reduces to the cubic equation

$$x^3 - 2dx^2 + d^2x - \frac{mR^3}{M} = 0,$$

and we look for a solution $x < R$. If, instead, for another example, we know a zero gravity point is inside m, we set the value of g_M *outside M* equal to the value of g_m *inside m*. Then,

$$\frac{GM}{x^2} = \frac{Gm}{r^3}(d-x),$$

which reduces to the cubic equation

$$x^3 - dx^2 + \frac{Mr^3}{m} = 0,$$

and we look for a solution $x > d - r$.

Here's an easy challenge problem. Show that if two massive spheres *just touch*, then the zero gravity point is at the point of contact *if* $\frac{r^2}{R^2} = \frac{m}{M}$. *Hint: Just touching* means $d = R + r$.

Here's a slightly more challenging challenge problem. Show that if $d = R\left(1 + \sqrt{\frac{m}{M}}\right)$ *and* if $d = r\left(1 + \sqrt{\frac{M}{m}}\right)$, then these two conditions together imply that the two spheres are just touching.

For a third challenge, suppose two new planets have been discovered, each with the same average density ρ as Earth. Let M, R, and g denote the Earth's mass, radius, and surface acceleration of gravity, respectively. If one of the new planets has twice the mass of Earth, and if the other new planet has twice the surface area of Earth, calculate the acceleration of gravity at the surface of each of the new planets. Now, with those last two calculations as warm-ups, here's a related question from a current (as I write) science story. In April 2017 astronomers announced the discovery of a rocky (that is, solid) planet orbiting a star in the constellation Cetus, about 39 light-years from Earth. This was exciting because the planet is in the star's so-called Goldilocks zone (that is, it is neither too hot nor too cold on the planet's surface, and so water can exist in the liquid state, a condition thought necessary for life to evolve). The planet has a diameter 40% larger than Earth's, and a mass 6.6 times that of Earth's. How much would a 150 pound astronaut weight on this planet?

Finally, here's a challenge question with a quite interesting practical conclusion. It is sometimes claimed that a journey from the Earth to the Moon is significantly aided by the "help" received from the Moon's gravity. Is that so? Here's a calculation to let you explore that claim. Since the radius of the Earth is 4,000 miles (call it R), and if we take the center-to-center distance between the Earth and the Moon as 240,000 miles, that distance is $60R$. As shown in the text, the location of Verne's point of zero gravity (in *From the Earth to the Moon*) is $0.9(60R) = 54R$ from the center of the Earth. Now, imagine a mass μ (Verne's manned capsule) transported from the Earth's surface ($x = R$) to the zero gravity point ($x = 54R$). Let E_1 be the energy required to make this journey against the gravitational pull of the Earth. Let E_2 be the energy *received* by μ from the Moon's gravitational pull during this journey. Calculate the numerical value of $\frac{E_2}{E_1}$. Use the fact that if the mass of the Earth is M, then the mass of the Moon is $\frac{1}{81}M$. *Hint* 1: The gravitational energy integrals are $\left| \int_{\text{start}}^{\text{finish}} F\,dx \right|$. That is, both E_1 and E_2 are positive. *Hint* 2: From the Moon's perspective, the journey (measured from the center of the Moon) is from $59R$ to $6R$. Given your result, how would you respond to the claim that the Moon's gravity is a significant aid in an Earth-to-Moon trip?

PROBLEM 21
Beyond the Cubic: Quartic Equations, Crossed Ladders, Undersea Rocket Launches, and Quintic Equations

Up to now all the algebraic equations we've encountered have been second-degree (quadratic) or third-degree (cubic). How about equations of the fourth degree (quartic)? What about a fifth-degree (quintic) equation? Does it take a weird, bizarre physical situation to have something like that appear in an analysis? Well, absolutely ... *not*. Let me start with a purely mathematical example, with a benign origin, in an ordinary physical situation that must occur daily *somewhere* on Earth. And then I'll show you more direct physical illustrations of how a quartic and a quintic can appear in a natural way.

Imagine two ladders, of lengths $m = 20$ feet and $n = 30$ feet, positioned so they cross each other across a lane of width w that is formed by two walls (see Figure P21.1). Each ladder reaches from the base of a wall to some point on the opposite wall. The point at which they cross is $H = 8$ feet above the ground. How wide is the lane (that is, what is w)?

To start our analysis, observe that we have two pairs of similar triangles: (ABE/ACD) and (BCE/ACF). This allows us to write, with the aid of the Pythagorean theorem,

$$\frac{a}{H} = \frac{w}{\sqrt{n^2 - w^2}},$$

and

$$\frac{b}{H} = \frac{w}{\sqrt{m^2 - w^2}}.$$

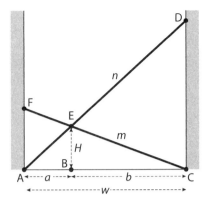

Figure P21.1. Two ladders across a lane.

Adding these two equations,

$$\frac{a}{H} + \frac{b}{H} = \frac{a+b}{H} = \frac{w}{H} = \frac{w}{\sqrt{n^2 - w^2}} + \frac{w}{\sqrt{m^2 - w^2}},$$

or

$$\frac{1}{H} = \frac{1}{\sqrt{n^2 - w^2}} + \frac{1}{\sqrt{m^2 - w^2}}.$$

Next, let's write y for the height at which the 20-foot ladder (m) touches its wall. That is, let's define

$$y = \sqrt{m^2 - w^2}$$

as the length AF. This means

$$w^2 = m^2 - y^2,$$

and so our previous equation becomes

$$\frac{1}{H} = \frac{1}{\sqrt{n^2 - m^2 + y^2}} + \frac{1}{y},$$

or

$$\sqrt{n^2 - m^2 + y^2} = \frac{1}{\frac{1}{H} - \frac{1}{y}} = \frac{Hy}{y - H}.$$

Now, squaring this and collecting terms, we get the quartic equation

$$y^4 - 2Hy^3 + \left(n^2 - m^2\right)y^2 - 2H\left(n^2 - m^2\right)y + H^2\left(n^2 - m^2\right) = 0.$$

Inserting the given values for m, n, and H, we have the perhaps scary-looking

$$y^4 - 16y^3 + 500y^2 - 8{,}000y + 32{,}000 = 0.$$

How do we solve something like this for y? *If* we can do that, *then w* immediately follows, since

$$w = \sqrt{m^2 - y^2} = \sqrt{400 - y^2}.$$

In fact, it is possible to find a formula for solving *any* quartic in terms of radicals involving just the coefficients of the equation, just as is the case for quadratic and cubic equations. Such a formula for quartics was found in 1540 by the Italian scholar Ludovico Ferrari (1522–1565).[1] I would be willing to make my standard $5 bet, however, that *no* mathematical physicist actually uses such a formula to solve a quartic he or she has just encountered; use of a computer is a far more attractive approach. Here's one way to do that.

From the boxed equation we have

$$\frac{1}{\sqrt{m^2 - w^2}} = \frac{1}{H} - \frac{1}{\sqrt{n^2 - w^2}},$$

---

[1] In 1826 Abel (see the beginning of Problem 12) showed that no such formula exists for algebraic equations of the fifth degree (quintic) or higher.

and so

$$\sqrt{m^2 - w^2} = \cfrac{1}{\cfrac{1}{H} - \cfrac{1}{\sqrt{n^2 - w^2}}},$$

or squaring both sides and solving for w in terms of itself,

$$w = \sqrt{m^2 - \left(\cfrac{1}{\cfrac{1}{H} - \cfrac{1}{\sqrt{n^2 - w^2}}}\right)^2}.$$

Now, we simply *iterate*. That is, we start with an initial guess for w (since clearly w is less than $m = 20$, I'll start with $w = 10$), plug that guess into the right-hand side of our last equation, and compute a new w on the left-hand side. Then, we keep repeating the process until we see our sequence of computed values of w converge to an obvious limit. In this case, we'll very quickly find that $w = 16.21$ feet.[2] You can verify this value by using it to calculate y (= 11.71 feet) and then plugging that value into the original y-quartic and seeing that it does, indeed, work.

Well, that problem was admittedly more mathematical than physical, so let me next show you how a quartic arises in what is undeniably a physical problem. With this problem we return (in spirit) to Problem 1, that of tossing a projectile, but now with a twist—the initial portion of the toss is underwater! With reference to Figure P21.2, here's what we have. A neutrally buoyant projectile is launched underwater from a compressed-air gun at a depth d, with an initial speed v_0, at an angle θ with respect to the horizontal. While moving upward toward the surface the projectile experiences a water drag force directly proportional to its instantaneous speed. What angle θ will give the projectile its maximum range R (measured from the launch point)?

By Archimedes' principle (derived in Problem 19, using a minimum-energy principle), a *neutrally buoyant* object displaces a volume of water of weight equal to the weight of the projectile on land. This means that if simply placed at some point underwater, the object will remain

---

[2] A little experimentation shows that the convergence is to this value for w *no matter what* value of w is used to start the iteration.

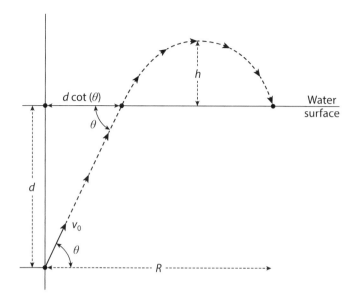

Figure P21.2. An underwater projectile launch.

at that point; the net force on it (gravity *downward* and buoyancy *upward*) is zero, as those two forces mutually cancel. (This is perhaps a somewhat artificial condition, but it will allow our analysis to be less messy than it would otherwise be.) Because the only force acting on the projectile during the underwater portion of its trajectory is water drag, acting in opposition to its motion, that motion will be a straight line, as shown in Figure P21.2 (h denotes the maximum height of the projectile). Once the projectile breaks through the water surface, however, then gravity *will* come into play.

Let m, v, and a denote the mass, the instantaneous speed, and the acceleration of the projectile, respectively, while underwater. Then, with c some constant, if we use s to measure the underwater distance along the straight-line path of the projectile, we can write

$$ma = m\frac{dv}{dt} = -cv,$$

where $-cv$ is the drag force (the minus sign is the *drag*). Since

$$v = \frac{ds}{dt},$$

then, by the chain rule of freshman differential calculus,

$$\frac{dv}{dt} = \frac{dv}{ds}\frac{ds}{dt} = v\frac{dv}{ds},$$

and so

$$mv\frac{dv}{ds} = -cv,$$

or

$$dv = -\frac{c}{m}ds.$$

This is easily integrated by noticing that $v = v_0$ when $s = 0$, and $v = v_1$ (the projectile speed as it breaks through the surface) when $s = \frac{d}{\sin(\theta)}$. So,

$$\int_{v_0}^{v_1} dv = -\frac{c}{m}\int_0^{\frac{d}{\sin(\theta)}} ds,$$

or

$$v_1 - v_0 = -\frac{c}{m}\frac{d}{\sin(\theta)}.$$

That is, we now have the launch of a projectile at initial speed v_1 at angle θ, where

$$v_1 = v_0 - \frac{c}{m}\frac{d}{\sin(\theta)}.$$

The subsequent motion of the projectile will, of course, now be a parabolic arc if we ignore (as we will) air drag.

From Figure P21.2 we can see that during the underwater motion, the projectile travels a horizontal distance $d\cot(\theta)$. Now, let's agree to measure time from $t = 0$ when the projectile emerges from the water at angle θ with horizontal and vertical speeds $v_1\cos(\theta)$ and $v_1\sin(\theta)$, respectively. So, the horizontal distance traveled by time t is $v_1 t\cos(\theta)$. The vertical speed at time t is $v_1\sin(\theta) - gt$ (where, of course, g is the acceleration of gravity). The projectile is at its maximum height h when the vertical speed is zero, and that occurs at time

$$t = \frac{v_1 \sin(\theta)}{g}.$$

So, by symmetry, the projectile hits the water at time

$$t = \frac{2v_1 \sin(\theta)}{g}$$

after traveling a horizontal distance of

$$v_1 \frac{2v_1 \sin(\theta)}{g} \cos(\theta) = \frac{2v_1^2}{g} \sin(\theta) \cos(\theta) = \frac{v_1^2}{g} \sin(2\theta).$$

Thus, the horizontal distance traveled *since launch* (that is, the range) is

$$R = d \, \cot(\theta) + \frac{v_1^2}{g} \sin(2\theta),$$

or, since

$$v_1 = v_0 - \frac{c}{m} \frac{d}{\sin(\theta)},$$

we have

$$R = d \, \cot(\theta) + \frac{1}{g} \left[v_0 - \frac{c}{m} \frac{d}{\sin(\theta)} \right]^2 \sin(2\theta).$$

To find the launch angle θ that maximizes R, we set $\frac{dR}{d\theta} = 0$. So,

$$-\frac{d}{\sin^2(\theta)} + \frac{1}{g} \left\{ \left[v_0 - \frac{c}{m} \frac{d}{\sin(\theta)} \right]^2 2 \cos(2\theta) \right.$$

$$+ \sin(2\theta) 2 \left[v_0 - \frac{c}{m} \frac{d}{\sin(\theta)} \right] \left[\frac{cdm \cos(\theta)}{m^2 \sin^2(\theta)} \right] \right\} = 0.$$

Now, we have $\cos(2\theta) = 1 - 2\sin^2(\theta)$ and $\sin(2\theta) = 2\sin(\theta)\cos(\theta)$, and so

$$\frac{gd}{\sin^2(\theta)} = \left[v_0 - \frac{c}{m}\frac{d}{\sin(\theta)}\right]^2 2\left[1 - 2\sin^2(\theta)\right]$$

$$+ 4\sin(\theta)\cos(\theta)\left[v_0 - \frac{c}{m}\frac{d}{\sin(\theta)}\right]\left[\frac{cd\cos(\theta)}{m\sin^2(\theta)}\right].$$

Thus,

$$gd = \left[v_0\sin(\theta) - \frac{cd}{m}\right]^2 2\left[1 - 2\sin^2(\theta)\right]$$

$$+ 4\sin(\theta)\left[v_0 - \frac{c}{m}\frac{d}{\sin(\theta)}\right]\left[\frac{cd\cos^2(\theta)}{m}\right],$$

or

$$gd = \left[v_0\sin(\theta) - \frac{cd}{m}\right]^2 2\left[1 - 2\sin^2(\theta)\right]$$

$$+ 4\left[v_0\sin(\theta) - \frac{cd}{m}\right]\frac{cd\left[1 - \sin^2(\theta)\right]}{m},$$

which, at this point, is clearly a quartic in $\sin(\theta)$. Perhaps this is as far as we need to go here (unless you are a *real glutton* for slogging through a mathematical jungle). To pursue the analysis further, this is the right place (in my opinion) to insert numerical values for g, d, c, v_0, and m and then solve the quartic *numerically* for $\sin(\theta)$ (and so for θ).

I'll let you play with that.

To study a physical system that gives rise to a quintic, we'll move into outer space. Imagine two massive bodies, one of which is much larger than the other (think of the Sun and the Earth), that are gravitationally interacting—for example, the Earth orbiting the Sun. It's easy to calculate the orbital period T of the Earth, as follows. Let's write the mass of the Sun as M_1, the mass of the Earth as M_2, and the distance between their centers of mass as R. The Earth's orbit is almost circular, and so if its orbital speed is v, then the centripetal acceleration of the Earth is

$$\frac{v^2}{R},$$

which means the force necessary to provide that acceleration is

$$M_2 \frac{v^2}{R}.$$

That force is supplied, of course, by the force of gravity, which is given by

$$G \frac{M_1 M_2}{R^2},$$

where G is the so-called gravitational constant (which we don't have to know for this analysis). So,

$$G \frac{M_1 M_2}{R^2} = M_2 \frac{v^2}{R},$$

which is easily solved to give Earth's orbital speed as

$$v = \sqrt{\frac{G M_1}{R}}.$$

Since the orbital path length is $2\pi R$, the orbital period immediately follows:

$$T = \frac{2\pi R}{v} = \frac{2\pi}{\sqrt{G M_1}} R^{3/2}.$$

Now, imagine a line joining the centers of the Earth and the Sun, as shown in Figure P21.3. The figure also shows a very small mass m (small in comparison to M_2, which in turn is small compared with M_1) also on the line joining M_1 and M_2, at distance r from M_2. Euler discovered in 1760 that there is a value for r so that the orbital period of m equals the orbital period of M_2. (Euler also found that there are two additional such points on the line joining M_1 and M_2—can you see

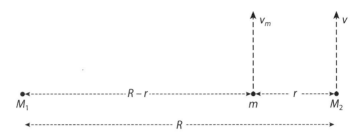

Figure P21.3. Mass m orbits M_1 in synchronism with M_2.

where they are?) A few years after Euler discovered his three points, the French-Italian mathematical physicist Joseph Lagrange (1736–1813) found (in 1772) two more such points *off* the M_1/M_2 line, for a total of five locations now collectively called the *Lagrangian points* (usually labeled L_1 to L_5). The mass m in Figure P21.3 is at L_1. Here's how to calculate r for L_1.

The net gravitational force on m, inward toward the Sun (M_1), is

$$\frac{GM_1m}{(R-r)^2} - \frac{GM_2m}{r^2} = Gm\frac{M_1r^2 - M_2(R-r)^2}{r^2(R-r)^2}.$$

Setting this force equal to the centripetal force on m, with v_m as the orbital speed of m, we have

$$m\frac{v_m^2}{(R-r)} = Gm\frac{M_1r^2 - M_2(R-r)^2}{r^2(R-r)^2},$$

which you can confirm gives

$$v_m = \sqrt{GM_1}\sqrt{\frac{r^2 - \frac{M_2}{M_1}(R-r)^2}{r^2(R-r)}}.$$

The orbital path length of m is $2\pi(R-r)$, and so if the orbital period of m is to equal that of the Earth, T, we have

$$\frac{2\pi(R-r)}{v_m} = \frac{2\pi R^{3/2}}{\sqrt{GM_1}},$$

which can be confirmed (with some easy algebra that I encourage you to do) to give

$$M_1 r^2 (R-r)^3 = r^2 R^3 M_1 - M_2 (R-r)^2 R^3.$$

By inspection, this is a fifth-degree equation in r, that is, a quintic, for which there is *no general formula* (see note 1 again).

In general, then, to solve for r, given the values of M_1 and M_2, requires a numerical attack, but in the case of $M_2 \ll M_1$ (the Earth/Sun case), we'd expect $r \ll R$. If you make that approximation, then it is not at all difficult to show that

$$r \approx R \left(\frac{M_2}{3M_1} \right)^{1/3}.$$

See if you can do this.

The Lagrangian point L_1 makes a dramatic appearance in the 2005 science fiction novel *Sunstorm*, by Arthur C. Clarke and Stephen Baxter, in which it is the location of a huge reflective shield, designed to save Earth from an enormous blast of energy from the Sun (a blast predicted to occur several years in the future). With a Sun/Earth mass ratio of 330,000 and $R = 93 \times 10^6$ miles, L_1 is

$$93 \times 10^6 \left(\frac{1}{990{,}000} \right)^{1/3} \approx 930{,}000 \text{ miles}$$

from the Earth's center. The novel says the location of L_1 is "about four times the distance to the Moon," which is correct (even though the Sun/Earth mass ratio is incorrectly stated to be 1 million, a value that is actually the *volume* ratio).

Now, here's a challenge problem for you. If a quartic has a special structure, then it sometimes can be solved analytically *without* the use of the general quartic formula. A trivial example of that would be $x^4 + ax^2 + b = 0$, which is simply a quadratic in thin disguise (let $y = x^2$). Much more interesting is something like $x^4 + ax^3 + bx^2 + ax + 1 = 0$. Find all four solutions in terms of a and b. *Hint* : Divide through by x^2 (which is okay to do because $x = 0$ is obviously *not* a solution, because $1 \neq 0$), and then let $y = x + \frac{1}{x}$.

Escaping an Atomic Explosion: Why the *Enola Gay* Survived

With all the higher-level math in the last few problems, are you a little nostalgic for the quadratic equation? If so, here's some relief. When the corrected paperback edition of my 2007 book *Chases and Escapes: The Mathematics of Pursuit and Evasion* (Princeton University Press) was published in 2012, it included an additional section not in the hardcover version. In that section I presented a discussion of how the famous World War II B-29 bomber *Enola Gay* escaped (just barely) the blast wave of the atomic bomb ("Little Boy") it had dropped less than two minutes earlier on Hiroshima, Japan, the morning of August 6, 1945.

As that discussion made clear, it wasn't at all obvious at the time that such an escape would even be possible. The men who flew history's first atomic bombing mission were taking a very big chance—and they knew it. What confidence they did have in the *possibility* they might survive was, to a large extent, a tribute to the solid reputation of the Ballistics Group of the Los Alamos Ordnance Division of the Manhattan Engineer District (the cover name for the American atomic bomb project). The Ballistics Group had carefully calculated what the *Enola Gay* should do after releasing the bomb; there were, of course, no absolute guarantees, but assuming it was even possible to fly out and away from a nuclear blast, the Ballistics Group's recommendations had a plausible chance of doing the job.

I won't repeat that detailed analysis here (it's on pages *xviii* to *xxvi* of the paperback edition of *Chases and Escapes*), but here's a summary of it. Flying directly in toward Hiroshima at an altitude of 31,600 feet at a speed of 328 miles per hour (481 feet per second), the *Enola Gay* released the bomb at time $t = 0$ (point A in Figure P22.1). Simultaneously with the weapon release, the pilot of the *Enola Gay* flipped the big plane almost literally over on its right wingtip and, diving slightly to increase speed to 350 miles per hour (513 feet per second), began executing a circular turn of radius $R = 4,720$

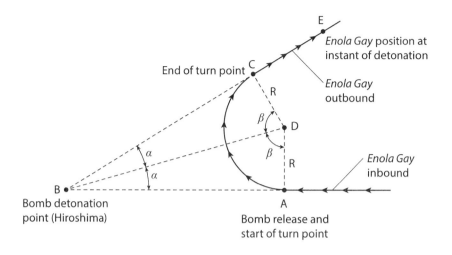

Figure P22.1. The escape of the *Enola Gay*.

feet, away from Hiroshima.[1] Meanwhile, the bomb continued on its downward, plunging path, falling directly toward Hiroshima until, at $t = 43$ seconds, it detonated at point B in Figure P22.1, at an altitude of 1,900 feet above the city. The ground distance covered by the falling weapon, from release at A to its explosion at B, was 19,600 feet. At the instant of the detonation, a supersonic shock wave traveling at 1,200 feet per second started on its way back toward the fleeing *Enola Gay*, which was at point E at the moment of detonation.

Obviously, the *Enola Gay*'s chances for surviving that shock wave would be best if the plane was as far away from the explosion as possible. Thus, the Ballistics Group told the pilot that after starting his turn to keep turning the plane until he reached point C, where the *Enola Gay* would be flying directly away from point B (that would be when the total turn angle, 2β in Figure P22.1, had reached 153°). At the instant he reached C, the pilot was to roll out of the turn and then continue to fly straight away from Hiroshima as fast as the *Enola Gay* could go ("throttles to the wall, pedal to the metal").

The analysis of this escape maneuver, as given in *Chases and Escapes*, calculated the slant range distance from the detonation point to the

---

[1] This perhaps curious value for the turning radius is developed in *Chases and Escapes* and is related to the g-forces experienced by the plane and its crew.

Enola Gay when the shock wave arrived as 63,600 feet, or about 12 miles, while the pilot later reported that "We were eleven and a half miles slant range from the atomic explosion."[2] Not too bad an agreement between theory and experiment.

Well, all that was a nice calculation, but once it was done and written up for the new edition of *Chases and Escapes* I moved on to other matters. That is, I did until when, in September 2015, I received an e-mail from Mr. M. R. Bhaskar, an aeronautical engineer in Hyderabad, India. In part, he wrote "I haven't understood why the escape trajectory [needs to be at 153°]. Assuming that the shock wave is spherically symmetric ... shouldn't a 90° turn work just as well?" I replied to that with a modified analysis that showed a 90° turn would have placed the *Enola Gay* 10,000 feet (in slant range) closer to the detonation at the instant the shock wave arrived.

But Mr. Bhaskar's e-mail did raise a quite interesting point: how do we *really* know that $2\beta = 153°$ was the best turning angle? It does have the psychologically pleasing feature, in the final leg of the escape, of having the plane flying directly away from the explosion, but might there be some other angle that would have achieved an even greater slant range separation? To study that question, consider Figure P22.2, which shows the *Enola Gay* turning through a general angle of θ and then flying along a straight path.

As in Figure P22.1, let's call the center of the turning circle D, which we'll set as the origin of a rectangular x,y-coordinate system. Thus the coordinates of D are $x = 0$, $y = 0$, and the coordinates of the weapon release point (A) are $x = 0$, $y = -R = -4,720$ (all distances are in feet) at time $t = 0$. Hiroshima is 1,900 feet below the detonation point of the bomb, which in turn is 29,700 feet below the horizontal plane that contains the flight path of the *Enola Gay*. When the bomb explodes at $t = 43$ seconds, point B (directly above the detonation point) in that horizontal plane has coordinates $x = -19,600$, $y = -4,720$. At a speed of 513 feet per second, the *Enola Gay* could complete one full circular

---

[2] Paul Tibbets, "How to Drop an Atom Bomb," *The Saturday Evening Post*, June 8, 1946.

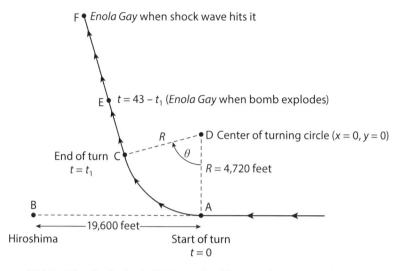

Figure P22.2. The *Enola Gay*'s flight path with a turning angle of θ radians (*not to scale*).

turn (2π radians) in

$$\frac{2\pi R}{\text{speed}}\text{seconds} = \frac{2\pi (4{,}720)}{513}\text{seconds,}$$

and so to turn through θ radians would take

$$\frac{\theta}{2\pi}\left(\frac{2\pi (4{,}720)}{513}\right) = \frac{4{,}720}{513}\theta\,\text{seconds} = t_1\,\text{seconds.}$$

Point C, the end of the turn, is therefore at $x = -4{,}720\sin(\theta)$, $y = -4{,}720\cos(\theta)$ at $t = t_1$. The *Enola Gay* then flies straight for $43 - t_1$ additional seconds at the speed of 513 feet per second, at which time the bomb explodes. Since the plane has flown a distance of $513(43 - t_1)$ beyond C, the Δx and Δy displacements from point C are

$$\Delta x = 513(43 - t_1)\cos(\theta),$$

and

$$\Delta y = 513(43 - t_1)\sin(\theta).$$

So, the coordinates of point E (where the *Enola Gay* is when the bomb explodes) are

$$x = -4{,}720 \sin(\theta) - 513(43 - t_1) \cos(\theta),$$

and

$$y = -4{,}720 \cos(\theta) + 513(43 - t_1) \sin(\theta).$$

The shock wave now starts on its way toward the fleeing *Enola Gay*.

T seconds after the bomb explodes, the shock wave hits the *Enola Gay* at point F. From point E to point F the *Enola Gay* flies an additional $513T$ feet, with Δx and Δy displacements from point E of

$$\Delta x = 513T \cos(\theta)$$

and

$$\Delta y = 513T \sin(\theta).$$

So, the coordinates of point F are

$$X = -4{,}720 \sin(\theta) - 513(43 - t_1 + T) \cos(\theta)$$

and

$$Y = -4{,}720 \cos(\theta) + 513(43 - t_1 + T) \sin(\theta).$$

Therefore, the slant range from detonation to where the shock wave hits the *Enola Gay* is, from the Pythagorean theorem, given by

$$\sqrt{(29{,}700)^2 + (X + 19{,}600)^2 + (Y + 4{,}720)^2}.$$

Now, we know that the slant range from B to F is also given by $1{,}200T$. So, for a given θ, we first calculate t_1 and then solve for T, such that

$$1{,}200T = \sqrt{(29{,}700)^2 + (X + 19{,}600)^2 + (Y + 4{,}720)^2}.$$

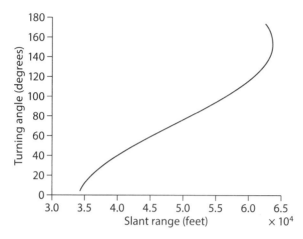

Figure P22.3. Finding the *Enola Gay*'s optimal turning angle.

In other words, we write

$$f = 1{,}200T - \sqrt{(29{,}700)^2 + (X + 19{,}600)^2 + (Y + 4{,}720)^2}$$

and find T so that $f = 0$. If we do this for lots of different values of θ (say $5° \leq \theta \leq 175°$ in steps of $0.1°$, a total of 1,701 values), a plot of the slant range ($1{,}200T$) versus θ will give us the optimal turning angle (the angle giving the largest slant range).

None of this is very complicated in theory, but a huge amount of numerical calculation is involved. Whatever the difficulty of doing that may or may not have been in 1945, today it's duck soup for even a run-of-the-mill laptop computer running any modern scientific programming language (like MATLAB). The code that does it all is, in fact, remarkably short (see **EG.m** and its subroutine **shock.m**), and it produced the plot shown in Figure P22.3. There we see that, indeed, $\theta \approx 150°$ produces a slant range separation of $\approx 63{,}000$ feet (just as hand calculated in *Chases and Escapes*), as well as demonstrating that the precise angle is *not* very critical; the maximum slant range is a quite broad maximum, and a turning angle anywhere in the interval $140° \leq \theta \leq 170°$ produces pretty nearly the same slant range separation between the detonation point and the *Enola Gay*'s location when the explosion shock wave arrives. To argue about whether the optimal turning angle is $153°$ or $158°$ (as has been done over the years) seems to be an argument about an insignificant point.

```
%EG.m/created by PJNahin for Mathematical Physics
AD=[5:0.1:175];
AR=AD*pi/180;
T1=4720*AR/513;
for loop=1:1701
    t1=T1(loop);
    angle=AR(loop);
    T=shock(t1,angle);
    SR(loop)=1200*T;
end
plot(SR,AD,'-k')
xlabel('slant range in feet')
ylabel('turning angle in degrees')
grid on
%shock.m/created by PJNahin for Mathematical
Physics (9/24/2016)
function delay=shock(t1,angle)
lower=0;upper=1000;
for k=1:15
    delay=(lower+upper)/2;
    term1=-4720*sin(angle)-513*(43-t1+delay)*cos(angle)+19600;
    term2=-4720*cos(angle)+513*(43-t1+delay)*sin(angle)+4720;
    f=1200*delay-sqrt(29700^2+term1^2+term2^2);
    if f>0
        upper=delay;
    else
        lower=delay;
    end
end
```

EG.m is a straightforward implementation of the equations in the text, with the line $T = shock(t1,angle)$ the only one perhaps needing some explanation. This is the line that, for the current values of θ (the value of *angle*) and t_1 (the value of *t1*), initiates the calculation of T (the time from detonation until the shock wave reaches the *Enola Gay*). This line passes the values of *angle* and *t1* to the subroutine **shock.m**, which then computes the value of *delay* (which ends up being assigned to T). The logic of **shock.m** is what computer scientists call the *binary chop algorithm*, whose great virtue is that it is guaranteed to converge. Here's how it works.

The variables *lower* and *upper* are initially set to values for T that are known to be *too small* and *too large*, respectively. That is, to 0 seconds and to 1,000 seconds. So, initially, we know that *f(lower)* < 0, and *f(upper)* > 0. Then, *delay* is recalculated as the average of *lower* and *upper* and, using that recalculated *delay*, f is calculated. If this $f > 0$, then *upper* is reduced to *delay*, while if $f < 0$, then *lower* is increased to *delay*. The difference between *upper* and *lower*, in any case, is cut in half. This is done a total of 15 times (a value picked somewhat at random), giving a total reduction by a factor of $2^{15} = 32,768$ of the initial difference between *upper* and *lower* of 1,000. That is, when **shock.m** returns to the main body of **EG.m**, the value of T has been estimated to within $\frac{1,000}{32,768} < \frac{1}{30}$ second. So, the value of $1,200T$, the slant range separation distance the shock wave has traveled from the detonation point until reaching the *Enola Gay*, has been found to within 40 feet. If that's not good enough, just let **shock.m** loop a few more times.

Okay, here's your assignment. Assume the *Enola Gay* bombing mission parameters are just as they were in 1945, with the exception of the bomb yield. In particular, suppose that instead of an average shock wave speed of 1,200 feet per second, the bomb was sufficiently more energetic that the average shock wave speed was doubled to 2,400 feet per second. How far away (slant range) would the *Enola Gay* have been when the shock wave hit, assuming the plane executed the optimal turn? What if the average shock wave speed was *tripled*, to 3,600 feet per second? The *Enola Gay* was severely shaken by the shock wave it actually experienced. Do you think it could have survived at the reduced distances you just calculated? Remember, the *Enola Gay* is now *closer* to a *bigger* explosion.

PROBLEM 23
Impossible Math Made Easy: Gauss's Congruence Arithmetic

Divide 5^{38} by 11. What's the remainder? If you are allowed to use a computer, then of course this is *technically* trivial. But also, of course, there are a *lot* of digits in 5^{38} (27, in fact, which is a lot more than a hand calculator can handle), and so you'll need a machine that can perform the calculation of 5^{38}, *exactly*, to answer the question. Amazingly, however, it is possible to answer this question with just a minute or so of extremely simple arithmetic *hand calculations*, ones that a bright high school student can learn to do by simply reading the rest of this discussion. It's all based on the theory of *congruences*, developed in 1801 by the great German mathematical physicist Carl F. Gauss (1777–1855).[1] Here's how it works.

Two integers, a and b, are said to be *congruent modulo m* if $a - b$ is evenly divisible by m. In other words, dividing a and b individually, each by m, produces the same remainder. If this is the case, we write

$$a \equiv b \,(\mathrm{mod}\, m).$$

So, for example, $9 \equiv 2(\mathrm{mod}\, 7)$ and $10 \equiv -1(\mathrm{mod}\, 11)$. This might all seem to be borderline trivial, but stay with me for just a bit longer. You'll change your mind! There are a number of fundamental theorems involving congruence arithmetic that are absolutely essential (and, fortunately, they are almost effortless to prove). They are:

Theorem A: If $a \equiv b(\mathrm{mod}\, m)$, then, with k any integer, $ak \equiv bk(\mathrm{mod}\, m)$;

Theorem B: If $a \equiv b(\mathrm{mod}\, m)$ and if $c \equiv d(\mathrm{mod}\, m)$, then $ac \equiv bd(\mathrm{mod}\, m)$;

Theorem C: If $a \equiv b(\mathrm{mod}\, m)$ and if $c \equiv d(\mathrm{mod}\, m)$, then $a + c \equiv b + d$ $(\mathrm{mod}\, m)$;

Theorem D: If $a \equiv b(\mathrm{mod}\, m)$ and if $b \equiv c(\mathrm{mod}\, m)$, then $a \equiv c(\mathrm{mod}\, m)$;

Theorem E: if $a \equiv b(\mathrm{mod}\, m)$ and if $ac \equiv d(\mathrm{mod}\, m)$, then $bc \equiv d(\mathrm{mod}\, m)$.

---

[1] Gauss was just 24 when he published his congruence arithmetic, in the first chapter of his masterpiece *Disquisitiones Arithmeticae* ("Arithmetical Investigations"). Gauss is one of history's three greatest mathematicians, on anybody's list (with Euler and Archimedes jostling with him for top spot), as well as being highly regarded in the world of physics. In the interest of history, though, I must tell you that a lot of the things in both math and physics that are named after Gauss were actually discovered by others. ("He that hath shall receive even more").

The first three theorems (A, B, and C) are just what we'd wish they'd be if we were hoping for theorems easy to remember; they simply say that congruence relations can be multiplied through by a constant, multiplied together, and added, just like ordinary equalities can. The fourth theorem (D) says congruence relations obey transitivity, while the fifth theorem (E) is just a bit more subtle in its meaning: if two numbers (a and b) are congruent, then we can replace either of them with the other in any other congruence (of the same modulo). I said all these theorems are easy to prove, and so let me now show you how true that is.

The proof of theorem A is almost trivial. To say $ak \equiv bk \pmod{m}$ is to say $ak - bk = (a - b)k$ is evenly divisible by m. Since we are given $a \equiv b \pmod{m}$—which means that $a - b$ is evenly divisible by m—then so is $(a - b)k$. DONE.

The proof of theorem B is only slightly more involved. To say $ac \equiv bd \pmod{m}$ is to say $ac - bd$ is evenly divisible by m. Since we are given $c \equiv d \pmod{m}$, we know that $c - d$ is evenly divisible by m, and so we can write $c - d = km$. That is, $c = d + km$, which means $ac - bd = a(d + km) - bd = ad + akm - bd = akm + (a - b)d$. Now, we are given $a \equiv b \pmod{m}$, which means $a - b$ is evenly divisible by m. Since akm is evenly divisible by m, then $ac - bd$ is evenly divisible by m. DONE.[2]

The truth of theorem C is equally easy to establish. The two given congruence relations say $a - b = km$ and $c - d = jm$ (where k and j are integers). Thus, $(a - b) + (c - d) = km + jm = (k + j)m = (a + c) - (b + d)$, which is simply a statement that $(a + c) - (b + d)$ is evenly divisible by m. That is, $a + c \equiv b + d \pmod{m}$. DONE.

For theorem D, we are given that $a - b = km$ and that $b - c = jm$. So, $b = c + jm$, which means that $a - (c + jm) = km$, or $a - c = jm + km = (j + k)m$. Thus, $a - c$ is evenly divisible by m, and so $a \equiv c \pmod{m}$. DONE.

Finally, here's the proof of theorem E. Writing $bc \equiv d \pmod{m}$ says $bc - d$ is evenly divisible by m. Is that true? We are given $ac \equiv d \pmod{m}$, and so $ac - d = jm$. We are also given $a \equiv b \pmod{m}$, which means $a - b = km$, or $a = b + km$. Thus, $ac - d = (b + km)c - d = jm$,

---

[2]Note, carefully, the following useful special case of $a = c$ and $b = d$. Then $ac \equiv bd \pmod{m}$ becomes $a^2 \equiv b^2 \pmod{m}$, and, in general, $a^k \equiv b^k \pmod{m}$ for k any positive integer. That is, we can raise a congruence to any positive power.

or $bc - d = jm - kmc = (j - kc)m$, and so $bc - d$ *is* evenly divisible by m. DONE.

Okay, we now have all we need to answer the question that opened this discussion. Prepare to be amazed!

We start with the observation

$$5^2 = 25 \equiv 3(\text{mod } 11).$$

How do I know to do this? Because I saw someone do it a long time ago when I was where you are right now, and now you can be me for the next person in the chain. How did the *first* person who did this make the key observation? He or she was *very* clever! Anyway, raising this congruence to the fifth power,

$$(5^2)^5 = 5^{10} \equiv 3^5(\text{mod } 11) = 9 \times 9 \times 3(\text{mod } 11)$$
$$= 243(\text{mod } 11) \equiv 1(\text{mod } 11);$$

that is,

$$5^{10} \equiv 1(\text{mod } 11).$$

Now,

$$5^{38} = 5^{30}5^8 = (5^{10})^3(5^2)^4,$$

which means (using theorems B and E)

$$5^{38} \equiv 1^3 3^4(\text{mod } 11) = 81(\text{mod } 11) \equiv 4(\text{mod } 11).$$

Thus, the remainder when dividing 5^{38} by 11 is 4. That's it!

We can use MATLAB to check our calculations, which tells us that[3]

$$5^{38} = 363,797,880,709,171,295,166,015,625.$$

---

[3]Writing $5 \wedge 38$ in **MATLAB** produces the double-precision result $3.637978807091713 \times 10^{26}$, which isn't precise enough for our question. Writing vpa($5 \wedge 38$), where vpa stands for "variable precision arithmetic" in **MATLAB**, will return the value 363797880709171286976233472, which is clearly still not good enough, because we know, *by inspection*, that the least significant digit is 5, not 2. The solution, in **MATLAB**, is to write vpa('$5 \wedge 38$'). My point here is that, whatever computer language your machine runs, you have to be alert for pitfalls like this and know how to avoid them.

If we then subtract 4 from this result, we find the result is evenly divisible by 11 (which shows that 5^{38}, when divided by 11, does indeed generate a remainder of 4).

The following is another example of a congruence calculation, one with an interesting story from the history of mathematics. A famous problem three centuries ago was that of proving Fermat's *conjecture* that $F_n = 2^{2^n} + 1$ is a prime for any integer $n \geq 0$. Over the years Fermat tried, numerous times, to prove his conjecture but never succeeded. The lure for him was that the first five cases *do* happen to be primes: 3, 5, 17, 257, and 65,537. Unknown to Fermat, however, is that the very next case, $F_5 = 2^{2^5} + 1 = 2^{32} + 1 = 4,294,967,297$, is divisible by 641 (which Euler discovered in 1732, and which explains Fermat's failure). This isn't really very difficult to show (any eighth-grader should be able to do the division!), but of course the more interesting question is how Euler arrived at even trying 641. Euler's analysis didn't use the congruence idea (because Gauss wouldn't be born until 45 years later) but, rather, some results from prime number theory. We, however, can use congruence theory to show that 641 divides $2^{32} + 1$ without having to calculate $2^{32} + 1$ or even doing a division.

The start of this analysis may seem to have little motivation but, once seen, it does begin to look "obvious." So,

$$640 = (64)(10) = (2^6)(5)(2) = 5(2^7).$$

Thus, since $640 \equiv -1 \pmod{641}$, we have

$$5(2^7) \equiv -1 \pmod{641}.$$

Then, raising both sides of this congruence to the fourth power, we have

$$5^4(2^{28}) \equiv (-1)^4 \pmod{641},$$

or

$$625(2^{28}) \equiv 1 \pmod{641}.$$

Next, we notice that $625 + 16 = 641$, and so

$$625 \equiv -16 \pmod{641},$$

or

$$625 \equiv -2^4 (\mathrm{mod}\, 641).$$

From theorem E, this means we can replace the 625 with -2^4 in $625(2^{28}) \equiv 1(\mathrm{mod}\, 641)$ to get

$$-2^4(2^{28}) \equiv 1(\mathrm{mod}\, 641),$$

or multiplying by -1 (using theorem A), we have

$$2^{32} \equiv -1(\mathrm{mod}\, 641),$$

which says $2^{32} + 1$ is divisible by 641. In fact, because I know you're curious,

$$\frac{2^{32} + 1}{641} = 6,700,417,$$

which is a prime (as is 641).

Here are three challenge questions for you.

(a) Prove that $2^{44} - 1$ is evenly divisible by 89. *Hint*: Start with $88 = 8 \times 11 = 2^3 \times 11$.

(b) Prove that $2^{48} - 1$ is evenly divisible by 97. *Hint*: Start with $96 = 32 \times 3 = 2^5 \times 3$.

(c) What is the remainder when you divide $1^5 + 2^5 + 3^5 + \ldots + 100^5$ by 4?

PROBLEM 24
Wizard Math: Fourier Series, Dirac's Impulse, and Euler's Zeta Function

One of the great intellectual achievements in 18th-century mathematics was the 1734 discovery by Euler (also a great mathematical *physicist*) that

$$\frac{1}{1^2} + \frac{1}{2^2} + \frac{1}{3^2} + \ldots = \sum_{n=1}^{\infty} \frac{1}{n^2} = \frac{\pi^2}{6} = 1.6449\ldots.$$

(You'll recall that I mentioned this famous result in Problem P3.) The value of the sum of the reciprocals of the integers squared had both fascinated and eluded mathematicians for centuries until Euler, using some highly controversial techniques (some of which were technically invalid!), finally did it. Today, many alternative, legal derivations are known, but they all still generally involve one or more ingeniously clever tricks. In fact, if we generalize Euler's problem just a bit, to read

$$\sum_{n=1}^{\infty} \frac{1}{n^s} = \zeta(s),$$

where $\zeta(s)$ is called the *Riemann zeta function*,[1] then Euler actually showed how to evaluate $\zeta(s)$ for all *even* integer values of s.[2] His 1734 result is for $s = 2$, and his next result for $s = 4$ is

$$\zeta(4) = \sum_{n=1}^{\infty} \frac{1}{n^4} = \frac{\pi^4}{90} = 1.0823\ldots.$$

One of the modern derivations for $\zeta(2)$ uses a clever integral approach that one mathematician said "provides a proof at *graduate level* [my emphasis] for Euler's $\zeta(2)$ result."[3] What I'll do next is show you an approach that all mathematical physicists (and engineers, too) are familiar with by the end of their second year of *undergraduate* studies, an approach that gives a *routine* way of calculating $\zeta(s)$ for *any* even value of s. I demonstrated this method for $\zeta(2)$ and $\zeta(4)$ in an earlier book,[4] so here I'll calculate $\zeta(6)$. Thus, there is no need to wait until *graduate school* (!) to see how to calculate $\zeta(2)$. Expressions for $\zeta(s)$ when s is *odd* remain, as in Euler's day, an immensely deep mystery. The first person to find a closed-form expression for $\zeta(3)$, similar to that for the $\zeta(s)$ when s is even, will go down in mathematical history as

---

[1]See Problem P3. For more on the Riemann zeta function and its connection to the most famous unsolved problem in mathematics today (the so-called Riemann hypothesis), see my book *Inside Interesting Integrals*, Springer, 2015.

[2]$\zeta(s) = \frac{a}{b}\pi^s$, where a and b are integers and s is even.

[3]Dirk Huylebrouck, "Similarities in Irrationality Proofs for π, ln2, $\zeta(2)$, and $\zeta(3)$," *American Mathematical Monthly*, March 2001, pp. 222–231.

[4]P. J. Nahin, *Mrs. Perkins's Electric Quilt*, Princeton University Press, 2009, pp. 94–102. In that book I discuss the occurrence of $\zeta(4)$ in a problem from solid-state physics.

a superstar, even if he or she never does anything else. Mathematicians are now beginning to suspect, however, that such expressions simply may not exist.[5]

One of the standard mathematical techniques all physicists (and engineers, too) are expected to know by the end of their sophomore year is *Fourier's theorem*: it is easy to state.[6] Suppose $f(x)$ is a function defined on the interval $-T \leq x \leq T$. Then, over that interval (and, actually, as an extended function over the entire infinite x-axis as a *periodic function* with period $2T$), we can write $f(x)$ in the form of an infinite sum of sine and cosine functions. That is, $f(x)$ is given by the *Fourier series*

$$f(x) = \frac{1}{2}a_0 + \sum_{n=1}^{\infty} \left\{ a_n \cos\left(\frac{n\pi x}{T}\right) + b_n \sin\left(\frac{n\pi x}{T}\right) \right\},$$

where the a_n and b_n coefficients[7] are given by

$$a_n = \frac{1}{T} \int_{-T}^{T} f(x) \cos\left(\frac{n\pi x}{T}\right) dx, \, n = 0, 1, 2, 3, \ldots ,$$

and

$$b_n = \frac{1}{T} \int_{-T}^{T} f(x) \sin\left(\frac{n\pi x}{T}\right) dx, \, n = 1, 2, 3, \ldots .$$

---

[5] Some things about $\zeta(s)$, for s odd, *have* been discovered since Euler's day. The French mathematician Roger Apéry (1916–1994) showed (in 1979) that $\zeta(3)$ (whatever its value may be) is irrational. In 2000, the French mathematician Tanguy Rivoal showed that, for an *infinity* of odd (but unspecified) values of s, $\zeta(s)$ is irrational. In 2001 he made that more specific by showing that there is at least one such odd s in the interval 5 to 21 (the Russian mathematician Wadim Zudilin later reduced the range on s to 5 to 11).

[6] If you want to brush up on the French mathematician Joseph Fourier (1768–1830) and his theorem, see my book *Dr. Euler's Fabulous Formula*, Princeton University Press, 2006, pp. 114–187.

[7] The physical meaning of the $\frac{1}{2}a_0$ term is that of the average value of $f(x)$ over the interval $-T \leq x \leq T$, that is, as the area bounded by $f(x)$ over that interval, divided by the length of that interval.

From this point on, I'll take $T = \pi$. That is, from now on we'll take

$$f(x) = \frac{1}{2}a_0 + \sum_{n=1}^{\infty} \{a_n \cos(nx) + b_n \sin(nx)\}, \; -\pi < x < \pi,$$

where the a_n and b_n coefficients are given by

$$a_n = \frac{1}{\pi} \int_{-\pi}^{\pi} f(x) \cos(nx) \, dx, \, n = 0, 1, 2, 3, \ldots,$$

and

$$b_n = \frac{1}{\pi} \int_{-\pi}^{\pi} f(x) \sin(nx) \, dx, \, n = 1, 2, 3, \ldots.$$

Now, before going any further, and to help convince you that this isn't all just "symbol pushing," let me show you an application of Fourier series that often occurs in mathematical physics. In the late 1920s, the great English physicist Paul Dirac (see note 4 in Problem 11) introduced the *unit impulse function* (or *Dirac delta function*) into quantum physics. This function, written as $\delta(x)$, is commonly thought of as being zero for all values of x *except* at $x = 0$, where it is infinite (that infinity is imagined to be such that the impulse bounds unit area). For a long time mathematicians thought Dirac's idea was crazy, declaring (in so many words) there is no such function. Just one of their objections to $\delta(x)$ is that despite being nonzero at just one point, it has infinite energy (more on that in just a bit). But, despite the criticism, Dirac replied by using his[8] function to solve problems that had stumped the mathematicians, which is actually a pretty good

---

[8]Dirac in fact (as he himself admitted) got the idea for the impulse from his study (as an undergraduate electrical engineering major) of the operational calculus developed decades earlier by the British electrical physicist Oliver Heaviside (who appears in Problem 26, and so I'll say no more about him, here). Electrical engineers found the impulse to be ideal in studying sudden transient phenomena (like a lightning bolt).

way to answer your critics. It's tough, after all, to complain very much about getting the right answer! The impulse proved to be so useful (and mathematicians eventually *did* develop a solid theoretical basis for it) that it is now a well-accepted part of mathematical physics.

A slightly less inflammatory way to define the impulse function is to say that given any continuous function $\phi(x)$, the impulse has the property that

$$\int_{-\infty}^{\infty} \delta(x) \phi(x) dx = \phi(0).$$

Using this *integral property* (sometimes called the *sampling property*) of the impulse, we can now calculate the Fourier series expansion for $\delta(x)$, one that is valid for all x in the interval $-\pi < x < \pi$, and in particular at $x = 0$, right where $\delta(x)$ "does its thing." To perform such a calculation might seem to be a major task, given the admittedly bizarre nature of $\delta(x)$, but actually it's a *trivial* task because of the integral property. To calculate a_n and b_n, all we have to do is write

$$a_n = \frac{1}{\pi} \int_{-\pi}^{\pi} \delta(x) \cos(nx) dx = \frac{1}{\pi} \cos(0) = \frac{1}{\pi}, \quad n = 0, 1, 2, 3, \ldots,$$

and

$$b_n = \frac{1}{\pi} \int_{-\pi}^{\pi} \delta(x) \sin(nx) dx = \frac{1}{\pi} \sin(0) = 0, \quad n = 1, 2, 3, \ldots.$$

Thus,

$$\boxed{\delta(x) = \frac{1}{2\pi} + \frac{1}{\pi} \sum_{n=1}^{\infty} \cos(nx), \quad -\pi < x < \pi.}$$

That's it; we're done!

But is the boxed expression correct? Figure P24.1 shows two MATLAB-generated plots of the right-hand side of the boxed expression, using the first 5 terms in the sum (on the left) and then the first 20 terms (on the right). You can literally see the impulse forming in these

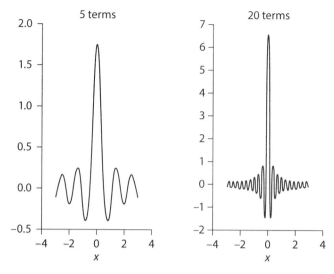

Figure P24.1. Two partial Fourier series expansions of $\delta(x)$.

plots, which is amazing in itself, and particularly so when you note the highly discontinuous nature of $\delta(x)$. What is happening is that the cosine terms are adding together (*constructively interfering*, in physics lingo) at $x = 0$, and *destructively interfering* at $x \neq 0$.

Earlier I told you that one of the objections to Dirac's impulse function is that it has infinite energy. Well, what does *that* mean? The terminology (used today by mathematicians as well as by physicists and engineers) comes from physics. If $x = t$ (that is, x is *time*), and if $f(t)$ is a voltage, then the integral

$$\int_{-\pi}^{\pi} f^2(t)dt$$

is the electrical energy delivered to a 1-ohm resistor over the time interval $-\pi < t < \pi$. So, the energy of the impulse is given by

$$\int_{-\pi}^{\pi} \delta^2(t)dt = ?$$

One way to attempt to "answer" this question is to write

$$\int_{-\pi}^{\pi} \delta^2(t)dt = \int_{-\pi}^{\pi} \delta(t)\delta(t)dt = \delta(0) = \infty,$$

where I've invoked the integral property, with one of the $\delta(t)$'s in the last integral playing the role of $\phi(t)$ (and then arguing that $\delta(0) = \infty$ because $t = 0$ is when the impulse becomes infinite). But remember, $\phi(t)$ was said to be a *continuous* function, and $\delta(t)$ is certainly *not* continuous. So this is really a pretty shaky line of reasoning. Is there a better way to show that $\delta(t)$ has infinite energy?

Yes, and Fourier series again comes to the rescue, in the form of *Parseval's theorem*:[9]

$$\frac{1}{2\pi} \int\limits_{-\pi}^{\pi} f^2(t)\,dt = \left(\frac{1}{2}a_0\right)^2 + \frac{1}{2}\sum_{n=1}^{\infty} (a_n^2 + b_n^2).$$

This is actually not difficult to establish, by writing

$$\frac{1}{2\pi} \int\limits_{-\pi}^{\pi} f^2(t)\,dt = \frac{1}{2\pi} \int\limits_{-\pi}^{\pi} \left[\frac{1}{2}a_0 + \sum_{n=1}^{\infty} \{a_n\cos(nt) + b_n\sin(nt)\}\right]$$

$$\times \left[\frac{1}{2}a_0 + \sum_{m=1}^{\infty} \{a_m\cos(mt) + b_m\sin(mt)\}\right] dt$$

and then multiplying out the right-hand side and integrating term by term.[10] The result is Parseval's theorem. For the impulse function we

---

[9]Named after the French mathematician Marc-Antoine Parseval des Chênes (1755–1836). The *entire* left-hand side of the theorem (with the $\frac{1}{2\pi}$ included) is called the *power* of $f(t)$, as it is the *energy* of $f(t)$ *divided by a time interval* of 2π (again, *physics* terminology).

[10]Using well-known trig identities for the products of the sine and cosine functions, it is easy to show the so-called *orthogonality conditions* that are quite helpful in doing the integrations:

$$\int\limits_{-\pi}^{\pi} \sin^2(nt)\,dt = \pi$$

$$\int\limits_{-\pi}^{\pi} \sin(nt)\sin(mt)\,dt = \int\limits_{-\pi}^{\pi} \cos(nt)\cos(mt)\,dt = 0,\ m \neq n$$

$$\int\limits_{-\pi}^{\pi} \cos^2(nt)\,dt = \begin{matrix} \pi, & n \neq 0 \\ 2\pi, & n = 0 \end{matrix}$$

$$\int\limits_{-\pi}^{\pi} \sin(nt)\cos(mt)\,dt = 0 \text{ for all } m, n.$$

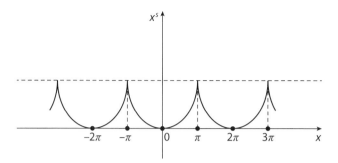

Figure P24.2. s even.

derived that $a_n = \frac{1}{\pi}$ for $n \geq 1$, and so, clearly,

$$\int\limits_{-\pi}^{\pi} \delta^2(t)dt = \infty,$$

and now we can *legitimately* claim that Dirac's impulse has infinite energy.

Okay, now back to the zeta function and how we can use Fourier series to evaluate it for even positive integer values of s. Suppose, in particular, that

$$f(x) = x^s, \; -\pi < x < \pi.$$

Then, $f(x)$, when *periodically extended* to the left and right of the $-\pi$ to π interval, looks like Figure P24.2 *because s is even*. In this case, the Fourier series converges, for every x, to the *continuous*, periodically extended $f(x)$. In particular, setting $x = \pi$ in the Fourier series will give the correct value of x^s. (Can you see what the periodically extended x^s looks like for s *odd*? What do you think the series converges to, then, at $x = \pi$?)

Let me now show you how to use a Fourier series to calculate $\zeta(6)$—and I think you'll be pleasantly surprised at just how fast (and easy) are the calculations are. We start by defining $f(x) = x^6$. Then,

$$a_0 = \frac{1}{\pi} \int_{-\pi}^{\pi} x^6 dx = \frac{1}{\pi} \left(\frac{1}{7} x^7\right) \Big|_{-\pi}^{\pi} = \frac{2}{7} \pi^6,$$

and[11]

$$a_0 = \frac{1}{\pi} \int_{-\pi}^{\pi} x^6 \cos(nx) dx = \frac{1}{\pi n^7} [6nx(n^4 x^4 - 20n^2 x^2 + 120)\cos(nx)$$

$$+ (n^6 x^6 - 30n^4 x^4 + 360n^2 x^2 - 720)\sin(nx)] \Big|_{-\pi}^{\pi}$$

$$= \frac{1}{\pi n^7} \left[12n^5 \pi^5 - 240n^3 \pi^3 + 1{,}440n\pi\right] \cos(n\pi)$$

$$= \left(\frac{12\pi^4}{n^2} - \frac{240\pi^2}{n^4} + \frac{1{,}440}{n^6}\right) \cos(n\pi).$$

We don't have to actually perform the b_n integration, because we know, *by inspection*, that the result is zero for all n. That's because $f(x) = x^6$ is an even function, and so the integrand of the b_n integral is an odd function ($\sin(nx)$ is odd, and even function times odd function is odd). And, of course, the integral of an odd function over an interval symmetrical about the origin vanishes. Thus,

$$x^6 = \frac{1}{7} \pi^6 + \sum_{n=1}^{\infty} \left(\frac{12\pi^4}{n^2} - \frac{240\pi^2}{n^4} + \frac{1{,}440}{n^6}\right) \cos(n\pi)\cos(nx).$$

In particular, setting $x = \pi$, this reduces to

$$\pi^6 = \frac{1}{7} \pi^6 + \left[12\pi^4 \sum_{n=1}^{\infty} \frac{1}{n^2} - 240\pi^2 \sum_{n=1}^{\infty} \frac{1}{n^4} + 1{,}440 \sum_{n=1}^{\infty} \frac{1}{n^6}\right],$$

or

$$\frac{6}{7} \pi^6 = 12\pi^4 \zeta(2) - 240\pi^2 \zeta(4) + 1{,}440 \zeta(6).$$

[11]I did *not* do the a_n integral by hand. Instead, the Mathematica integrator available (for free) on the Web, from Wolfram, is the hero.

Substituting the known values of $\zeta(2)$ and $\zeta(4)$, we obtain

$$\frac{6}{7}\pi^6 = 12\pi^4\frac{\pi^2}{6} - 240\pi^2\frac{\pi^4}{90} + 1{,}440\zeta(6),$$

or

$$\pi^6\left(\frac{6}{7} - 2 + \frac{240}{90}\right) = 1{,}440\zeta(6),$$

or, finally,

$$\zeta(6) = \frac{\frac{6}{7} - 2 + \frac{8}{3}}{1{,}440}\pi^6 = \frac{\pi^6}{945} = 1.0173\ldots.$$

Here are three challenge questions for you.

(a) Comment on the physical significance of the $\frac{1}{2\pi}$ term in the boxed Fourier series for $\delta(x)$. You'll recall (note 7) that I said the constant term in *any* Fourier series is the average value of $f(x)$ over $-\pi < x < \pi$, but how do you average a function like $\delta(x)$ that is zero everywhere except at *one point* $(x = 0)$, where it is infinite? How do we get $\frac{1}{2\pi}$ from that?

(b) Consider the periodic "square-wave" function shown in Figure P24.3. Calculate the Fourier series for that function, and then, using Parseval's theorem, calculate the *exact* value of

$$\sum_{n\,\text{odd}}\frac{1}{n^2} = 1 + \frac{1}{3^2} + \frac{1}{5^2} + \frac{1}{7^2} + \cdots.$$

Then, show how this can be used to derive Euler's result for $\zeta(2)$.

(c) Find the Fourier series expansion for $f(x) = x^2$, $-\pi < x < \pi$, and then integrate it indefinitely to find the *exact* value of

$$\frac{1}{1^3} - \frac{1}{3^3} + \frac{1}{5^3} - \frac{1}{7^3} + \cdots,$$

a result first calculated (using a different approach) by Euler. Notice the similarity of this expression to $\zeta(3)$, for which the exact value remains a mystery. As I write (in 2017), the *numerical* value of $\zeta(3)$ has been

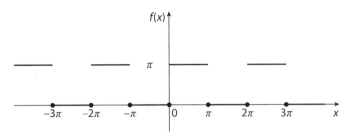

Figure P24.3. A square-wave.

computed out to 400 billion decimal digits, but that simply adds to the frustration of the failure, so far, by mathematicians to do for $\zeta(3)$ what Euler did for $\zeta(2n)$ *centuries* ago. If the number gods are anything, they are perverse.

(d) It can be shown that the Fourier series expansion of the function

$$f(x) = -\ln\left|2\sin\left(\frac{x}{2}\right)\right|, -\pi \le x \le \pi$$

is[12]

$$f(x) = \cos(x) + \frac{\cos(2x)}{2} + \frac{\cos(3x)}{3} + \frac{\cos(4x)}{4} + \frac{\cos(5x)}{5} + \cdots.$$

Use this and Parseval's theorem to evaluate

$$\int_{0}^{\pi} f^2(x)\,dx = \int_{0}^{\pi} \ln^2\left\{2\sin\left(\frac{x}{2}\right)\right\}dx,$$

an integral that might otherwise be pretty tough to do. (Notice that we can drop the absolute value signs in the integrand because, over the interval of integration, $2\sin\left(\frac{x}{2}\right)$ is nonnegative.)

---

[12]See, for example, Georgi P. Tolstov, *Fourier Series*, Dover, 1962, pp. 91–93. The derivation of this series is all freshman calculus, but is just a bit too long to include here. Tolstov (1911–1981), a well-known mathematician at Moscow State University, was the author of numerous acclaimed math books. His particularly beautiful *Fourier Series* should, I think, be in the personal library of all mathematical physicists.

PROBLEM 25
The Euclidean Algorithm: The Zeta Function and Computer Science

As shown in the previous discussion, the zeta function, despite its perhaps esoteric appearance, does have interest for physicists and electrical engineers and, perhaps even more surprising, for computer scientists, too. Here's an example of that. Modern-day data encryption systems depend on the inherent difficulty in factoring a very large number (one with, say, *hundreds* of digits) into a product of primes. The underlying mathematics of these systems then quickly leads to the consideration of what are called *relatively prime* (or *coprime*) integers. (You won't immediately see how the zeta function appears, but hang in there, and I think your patience will be rewarded.)

Two integers, even if composite, are said to be *relatively prime* if they have no common factors. For example, neither 8 or 9 is prime, but they are *relatively* prime, as $8 = 2 \times 2 \times 2$, while $9 = 3 \times 3$. However, 9 and 12 are *not* relatively prime, as $12 = 2 \times 2 \times 3$, and so (obviously) 9 and 12 share the factor 3. Another way of expressing this is to say that the *greatest common divisor* (gcd) of relatively prime numbers is 1. Being able to efficiently compute the gcd of two arbitrary integers proves to be central to some very important, practical problems, such as data encryption. So, I'll next show you the commonly used algorithm[1] for calculating the gcd, which requires knowing only *how to divide*.

That last assertion is a really amazing claim, so let me say it again: the Euclidean algorithm lets you calculate the gcd of two integers *without having to factor either one*. This is one of those infrequent times when something "too good to be true" is actually true. As a prelude to discussing that important algorithm, let me first establish a preliminary gcd result that we'll need.

Theorem: If a and b are positive integers such that $a \geq b > 0$—which means $a = qb + r$, where $q \geq 1$ and $0 \leq r < b$ (by what mathe-

---

[1]Called the *Euclidean algorithm* because it appears in Euclid's *Elements* (historians of mathematics, however, believe it is actually even older than that).

maticians call the *division algorithm*, which I'll take as being obvious)— then $\gcd(a,b) = \gcd(b,r)$.

Proof: Let's write $d = \gcd(a,b)$, which means d is the largest integer that divides *both* a and b. Thus, since $r = a - qb$, it is clear that d also divides r. Now, let c be *any* common divisor of b and r, which means c certainly divides $a = qb + r$. Thus, c is a divisor of a and b, just as d is (since c is any common divisor, perhaps $c = d$). But since d is the *largest* common divisor of a and b, it must be true that $c \leq d$. Therefore, $\gcd(a,b) = \gcd(b,r)$, and we are done. Now we can state the following:

Euclidean Algorithm

Let a and b be two positive integers, with $a \geq b > 0$. Then, by the division algorithm, we can write

$$a = q_1 b + r_1,$$

where $q_1 \geq 1$ and $0 \leq r_1 < b$. If $r_1 = 0$, then b divides a (and of course b divides b), and so we immediately have $b = \gcd(a,b)$. But what if $r_1 \neq 0$? Then we write (by the division algorithm)

$$b = q_2 r_1 + r_2,$$

where $q_2 \geq 1$ and $0 \leq r_2 < r_1$. If $r_2 = 0$, then stop. If $r_2 \neq 0$, then write (by the division algorithm)

$$r_1 = q_3 r_2 + r_3,$$

where $q_3 \geq 1$ and $0 \leq r_3 < r_2$. If $r_3 = 0$, then stop. If $r_3 \neq 0$, then write (by the division algorithm)

$$r_2 = q_4 r_3 + r_4,$$

where $q_4 \geq 1$ and $0 \leq r_4 < r_3$. If $r_4 = 0$, then stop. If $r_4 \neq 0$, then write (by the division algorithm) ... and so on in this way until you *do* get a zero remainder. Let's say this happens at the $(n+1)$th stage, and so we have for our final two equations

$$r_{n-2} = q_n r_{n-1} + r_n,$$

where $q_n \geq 1$ and $0 \leq r_n < r_{n-1}$, and

$$r_{n-1} = q_{n+1}r_n.$$

The final equation is *guaranteed* to occur (that is, we are *certain* to eventually get a zero remainder) because the sequence of remainders is continuously decreasing ($r_1 > r_2 > \ldots > r_n > 0$), and any sequence of continually decreasing, nonnegative integers is bound to reach zero in a finite number of steps. Now, the claim is that the last nonzero remainder, r_n, is, the gcd(a,b). Before proving this, here's an example of the Euclidean algorithm in action, finding the gcd of 161 and 713:

$$713 = 4(161) + 69$$
$$161 = 2(69) + 23$$
$$69 = 3(23) + 0$$

and so gcd($713,161$) = 23. This is an amazing calculation, as we have found the greatest common factor (divisor) of two numbers *without* having to actually factor either one. The Euclidean algorithm is very easy to code for a computer, and here's how I coded it in MATLAB on my laptop:

```
%gcd.m/created by PJNahin for Mathematical Physics
a=713;b=161;
r=1;
while r>0
    q=floor(a/b);
    r=a-q*b;
    a=b;
    b=r;
end
a
```

When run, **gcd.m** calculated gcd(713,161) = 23 in less than ten *microseconds*. And just to see what happens with larger numbers, see if you can show, using pencil and paper, that gcd(9767666321, 8876532413) = 7. (This again took **gcd.m** only microseconds to do—how long did it take *you?*) I'll show you, in just a moment, why **gcd.m** is so fast. But first, let's prove that $r_n = \gcd(a,b)$.

By our theorem we have

$$\gcd(a, b) = \gcd(b, r_1).$$

By the same theorem we also have

$$\gcd(b, r_1) = \gcd(r_1, r_2),$$

and

$$\gcd(r_1, r_2) = \gcd(r_2, r_3),$$

as well as

$$\gcd(r_2, r_3) = \gcd(r_3, r_4),$$

and so on, all the way to

$$\gcd(r_{n-2}, r_{n-1}) = \gcd(r_{n-1}, r_n),$$

and finally

$$\gcd(r_{n-1}, r_n) = \gcd(r_n, 0) = r_n.$$

So, stringing all these statements together, we have

$$\gcd(a, b) = \gcd(b, r_1) = \gcd(r_1, r_2) = \gcd(r_2, r_3) = \ldots = r_n,$$

which establishes the Euclidean algorithm, and we are done.

The Euclidean algorithm is very fast, and it's easy to show why. The speed is due to two reasons: (1) each step involves only a single division, and (2) the number of divisions required increases only as the *logarithm* of the product ab (the logarithm is a slowly increasing function of its argument). Specifically, what I'll show next is that an upper bound on the number of divisions required is given by

$\log_2(ab)$. For example, with $a = 713$ and $b = 161$ this upper bound is $\log_2(114{,}793) = 17$ divisions, which is significantly greater than the actual value of 3 divisions (that's why the theoretical expression is called an *upper* bound).[2] Here's how to derive the $\log_2(ab)$ upper bound (and going through this derivation will help you, at least as much as did the formal proof, in *understanding* the algorithm).

When we start the Euclidean algorithm with the two numbers a and b, their product is given by (of course!) ab. The first equation is given by

$$a = bq_1 + r_1,$$

where $0 \le r_1 < b$, and the second equation is given by

$$b = r_1 q_2 + r_2.$$

Now, think of the second equation as the starting equation for the two numbers b and r_1, with the product br_1. How does br_1 compare with the initial product ab? To answer this, notice that b is either less than $\frac{1}{2}a$, or it is at least as large as $\frac{1}{2}a$. Suppose it's the first case; that is, $b < \frac{1}{2}a$. Since $r_1 < b$, then it's certainly true that $r_1 < \frac{1}{2}a$ if $b < \frac{1}{2}a$. Suppose, however, it's the second case; that is, $b \ge \frac{1}{2}a$. Then, as

$$r_1 = a - bq_1,$$

and since $q_1 \ge 1$, then it's certainly true that $r_1 < a - b$, as the right-hand side is as large as it can be (using $q_1 = 1$). Then, using the smallest value for b to make the right-hand side as large as possible, we have

$$r_1 < a - \frac{1}{2}a = \frac{1}{2}a,$$

---

[2] In 1844 the French mathematician Gabriel Lamé (1795–1870) proved that another upper bound on the number of divisions is five times the number of digits in the smaller of a and b, and so for our example (since 713 has three digits) a slightly smaller upper bound is 15 divisions.

the same result we got for the first case. That is, *no matter what b is,* $r_1 < \frac{1}{2}a$. So, multiplying through by b,

$$0 \leq br_1 < \frac{1}{2}ab.$$

If we do all this again for the next equation in the Euclidean algorithm we find that

$$0 \leq r_1r_2 < \left(\frac{1}{2}\right)^2 ab.$$

That is, each new equation has a "product" that is the previous product divided by at least 2. So, when we get to r_n, we have

$$0 \leq r_{n-1}r_n < \left(\frac{1}{2}\right)^n ab.$$

This division cannot, of course, go on forever, as a decreasing sequence of positive integers (as is the sequence of products) must terminate once

$$\left(\frac{1}{2}\right)^n ab < 1,$$

as 1 is the smallest of the positive integers. When does this happen? That is, what is n, the number of divisions after which the Euclidean algorithm *must* terminate?

Taking logarithms to the base 2, and asking what n is when

$$\left(\frac{1}{2}\right)^n ab = 1,$$

we have

$$\log_2 \left\{ \left(\frac{1}{2}\right)^n ab \right\} = \log_2 \left\{ \left(\frac{1}{2}\right)^n \right\} + \log_2 \{ab\} = \log_2 \{1\} = 0,$$

or

$$\log_2\left\{\frac{1}{2^n}\right\} = -\log_2\{ab\} = \log_2\{1\} - \log_2\{2^n\} = 0 - n\log_2\{2\} = -n,$$

or, at last,

$$n = \log_2\{ab\},$$

just as claimed. To take a dramatic example from data encryption schemes, if both a and b are monster 300-digit numbers, then a and b are each at most 10^{300}, and so calculating their gcd would require no more than $\log_2(10^{300}10^{300}) = \log_2(10^{600}) = 600\log_2(10) < 2,000$ divisions. This would of course be a terrible task for a human, but a modern computer could blast through the calculations in far less than a second.

Now, as part of the theory of encryption systems, the following question soon arises: if we select two integers a and b at random from all the integers from 1 to n, what's the probability that a and b are relatively prime (there is no prime factor common to a and b)? A theoretical analysis[3] leads to the (surely surprising) result that in the limit as $n \to \infty$, the probability is

$$\frac{1}{\zeta(2)} = \frac{6}{\pi^2} = 0.6079\ldots.$$

To see the zeta function occurring in this problem *is* surprising— perhaps, in fact, surprising to the point that you might wonder if there is some way to at least partially "verify" this claim. By now you surely must have guessed *my* answer to that: let's use our favorite software to *simulate* the problem!

By that I mean, for a given value of n, let's pick the integers a and b at random from 1 to n. Then, once we have a and b, we'll run them through the Euclidean algorithm. In fact, we'll do that many times, say *10 million times*, because the Euclidean algorithm is so very fast, and

---

[3] Take a look, for example, at an elegant book by Julian Havil, *Gamma: Exploring Euler's Constant*, Princeton University Press, 2003, pp. 65–68.

see how many of those times gcd(a,b) = 1. If the theory is correct as we do this with increasing values of n, we should see the simulation results for the probability of a and b being coprime approach 0.6079. This is a straightforward simulation problem, and the MATLAB code **coprime.m** does the job (**coprime.m** is simply **gcd.m** embedded in a loop). Here are the results:

n	Probability that a and b are relatively prime
100	0.6087
1,000	0.6084
10,000	0.6079
100,000	0.6080

Well, I'm convinced!

```
%coprime.m/created by PJNahin for Mathematical Physics
sum=0;
n=input('What is n?')
for loop=1:10000000
    a=floor(n*rand)+1;
    b=floor(n*rand)+1;
    r=1;
    while r>0
        q=floor(a/b);
        r=a-q*b;
        a=b;
        b=r;
    end
    if a==1
        sum=sum+1;
    end
end
sum/loop
```

As a final comment on the zeta function and relatively prime numbers, it turns out that the result generalizes in a natural way: the probability that k integers, each selected at random from 1 to n, are relatively prime is, *as a group* (as $n \to \infty$), given by $\frac{1}{\zeta(k)}$. A subtle complication to simulating this generalization is that it is possible to have k relatively prime integers that are *not* relatively prime when taken in pairs. For example, $6(= 3 \times 2)$, $10(= 5 \times 2)$, and $15(= 5 \times 3)$ are not relatively prime in pairs but *are* relatively prime as a triplet. So, how do we handle this situation?

Theory predicts, for example, that three randomly selected integers from the interval 1 to n will as $n \to \infty$ be relatively prime as a group with probability

$$\frac{1}{\zeta(3)} = \frac{1}{1.2020569\ldots} = 0.8319.$$

How can we "confirm" this with a simulation? The key idea is to use the fact that if a, b, and c are the three integers, and if we write $d = \gcd(a, b, c)$, then

$$d = \gcd(\gcd(a, b), c).$$

This is not a particularly difficult result to establish (for now, I'll simply assume it's true), and it tells us that we can use the Euclidean algorithm, first, on a and b to find $f = \gcd(a, b)$ and then again on f and c to find d. A MATLAB code that does this, **triple.m**, is simply **coprime.m** extended in the obvious way.

```
%triple.m/created by PJNahin for Mathematical Physics
sum=0;
n=input('What is n?');
for loop=1:10000000
    a=floor(n*rand)+1;
    b=floor(n*rand)+1;
```

(continued)

(continued)

```
c=floor(n*rand)+1;
r=1;
while r>0
    q=floor(a/b);
    r=a-q*b;
    a=b;
    b=r;
end
b=c;
r=1;
while r>0
    q=floor(a/b);
    r=a-q*b;
    a=b;
    b=r;
end
if a==1
    sum=sum+1;
end
end
sum/loop
```

The results are *highly* convincing:

n	Probability that a, b, and c are relatively prime
100	0.8326
1,000	0.8322
10,000	0.8320
100,000	0.8319

For your challenge, prove that $\gcd(a, b, c) = \gcd(\gcd(a, b), c)$.

PROBLEM 26
One Last Quadratic: Heaviside Locates an Underwater Fish Bite!

The start of this book was prompted by a *Boston Globe* letter to the editor that took a jaundiced view of the practical value of knowing how to solve quadratic equations. That did get the book off the ground, but all the while I wrote a nagging question bedeviled me: how to *end* it? Physics finally gave me the answer, an answer so obvious I now wonder why it took me so long to see it. *Symmetry* is a fundamental concept for mathematical physicists, and so ending the book with a return to the quadratic equation was clearly the correct path to follow.

One of the most important (and, without exaggeration, *fantastic*) achievements of electrical and mechanical engineering, and of physics, in the 19th century was the beginning of the laying of underwater telegraph cables in the 1850s, cables that eventually linked the land-lines in England and Europe with those in America. (The most famous of all such cables is almost surely the nearly 1,700-mile-long trans-Atlantic cable of 1865–1866, weighing 5,000 tons.[1]) Telegraph cables are obviously electrical in nature, and so the *electrical* engineering involved in such a cable is obvious, as is the *mechanical* engineering aspect; after all, the task of laying a massive cable weighing tons per mile off the back of a rolling ship, into water thousands of feet deep, is not that of just dumping it overboard.

The physics aspect of undersea cables is perhaps not so obvious. The original electrical theory behind such cables is due to the Scottish scientist William Thomson (1824–1907), better known in the world of physics as Lord Kelvin. His theory *was* later shown to have some serious deficiencies, but it was nevertheless sufficiently correct that it made the *possibility* of a long undersea cable a reasonable consideration. When such cables were actually constructed and put into commercial

---

[1]The amazing story behind this achievement is well told by Bern Dibner, *The Atlantic Cable*, Burndy Library, 1959. And for a literally eyewitness, on-the-scene account, see W. H. Russell, *The Atlantic Telegraph*, Nonsuch Publishing, 2005 (original published in 1866). William Howard Russell (1820–1907) was a reporter for *The Times* who sailed on the *Great Eastern* when it laid the Atlantic cable.

operation, the achievement was correctly considered to be only slightly less amazing than a miracle.

Suddenly, you see, the time required to send a message 2,000 miles and more across the Atlantic Ocean went from weeks in a sealed mailbag on a ship (with a not-insignificant chance of sinking), to mere minutes as electrical impulses traveling at speeds measured in many miles per second. The result was a revolution in communication that wouldn't be equaled until more than half a century later, with the development of broadcast radio. This revolution in the near-instantaneous connection of widely separated parts of the world has rightly been called the "Victorian Internet."[2]

But, of course, submerged cables had their operational problems, too. The central one was that, now and then, they developed breaks. That is, the insulating sheath surrounding the central electrical wires would, for a variety of reasons, develop an electrical "leak" to the surrounding seawater, forming a finite resistance path from the interior of the cable to what could be taken, as a first approximation, to be an electrical *ground* (a point at zero volts). When such a break occurred, the reception of a message at the distant end of the cable was greatly impaired, or even rendered impossible. One obvious reason for the development of a break was, of course, simply that of mechanical abrasion. That is, shifting water currents would move the cable back and forth across a rough sea bottom and thereby tear open the cable's protective shielding.

Another, less obvious cause for a break was a fish bite! As one correspondent noted in an 1881 letter to the English trade journal *The Electrician*, a cable laid in 1874 was soon after found to have suffered "at least four indubitable fish-bite faults ...where the iron sheathing had been forcibly crushed up and distorted ...as if by the powerful jaws of some marine animal." That writer presented convincing evidence pointing to *Plagyodus ferox* ("one of the most formidable of deep-sea fishes") as the culprit.

Now, one could at least *imagine* a repair ship sailing along the known path of the submerged cable, periodically grabbling for the cable, and

---

[2]Tom Standage, *The Victorian Internet: The Remarkable Story of the Telegraph and the Nineteenth Century's On-line Pioneers*, Walker, 1998.

then pulling it up to the surface to see if the damaged portion had been found. Considering the lengths of cables, and the depths at which they lay, however, it is easy to see that this could be a lengthy and *very* expensive approach, similar to searching for a needle in a haystack by repeatedly sticking your hand into the hay until you stabbed yourself. A much better approach uses a technique called the *Blavier method,*[3] and it all depends on knowing how to solve a quadratic equation.

Whatever the cause of a break, before a damaged cable would completely separate it would first develop what is called a *fault,* an electrical path of some finite resistance from its still physically intact internal wires to the surrounding water. While the original electrical circuit was now degraded, often greatly so, it was still a *circuit* and that was the key to being able to pinpoint the location of the fault, thus allowing the direct recovery and repair of the cable.

We'll mathematically model the cable, and its fault, as shown in Figure P26.1. We'll assume the cable has a constant, known electrical resistance per mile (measured at the time of the cable's construction), and that the total resistance of the originally intact cable, from one end to the other, is a ohms.[4] Further, we'll write the resistance of the cable from its left end to the fault as x ohms, and so the resistance from the fault to the right end is $a - x$ ohms. And finally, the resistance between the fault and the surrounding seawater will be taken to be y ohms. If we can determine the value of x, then we can use the resistance per mile of the cable to calculate how far the fault is from either end.

Blavier's method requires two separate electrical measurements. First, with the right end of the cable *not connected to anything*, the resistance "seen looking into" the left end was measured.[5] That is, a

---

[3]After its inventor, the French telegraph engineer Edouard Ernst Blavier (1826–1887).

[4]The *ohm,* named after the German mathematical physicist Georg Simon Ohm (1789–1854), is an MKS unit (see Appendix 4).

[5]For a cable hundreds of miles long, how would the operator at the left end know when the operator at the right end had disconnected? One way to conceptualize how that could be accomplished is to simply schedule the Blavier test to be performed at regular times, with the two operators equipped with synchronized clocks. Notice that *if there were no fault* (that is, if $y = \infty$), then $b = \infty$ and $c = a$, which would mean all is well.

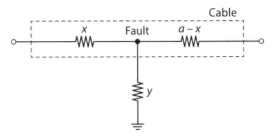

Figure P26.1. Simple Blavier circuit for a cable fault.

known voltage would be applied to the cable end and the resulting current measured. By Ohm's law, the resistance would be recorded as b, where

$$b = x + y,$$

because x and y are in series, and series-connected resistors add. Next, the right end of the cable would be connected directly to ground (thus placing $a - x$ in parallel with y), and the left-end resistance measurement repeated. By Ohm's law this resistance would be recorded as c, where

$$c = x + \frac{y(a - x)}{y + (a - x)},$$

because x and the parallel combination of $a - x$ and y are in series, and parallel resistances combine as their product divided by their sum.

We now have two equations in two unknowns (x and y), and we want to solve for x. This is easy to do. Since $y = b - x$, then substituting that expression into the equation for c gives, with very little algebra, the quadratic equation

$$x^2 - 2cx + c(a + b) - ab = 0,$$

which is quickly solved with the aid of the quadratic formula (and so we again see that such knowledge isn't quite as useless for practical work as implied by the *Boston Globe* letter), to give us *two* real values for x:

$$x = c \pm \sqrt{(a - c)(b - c)}.$$

They are *real* because the square root is certain to be real, as both $(a - c)$ and $(b - c)$ are positive quantities.

To see the truth of that last statement, first notice that for the $(a - c)$ factor we can argue that $c < a$ because $a - x$, in parallel with y, will clearly be less than $a - x$. And for the $(b - c)$ factor we have

$$b - c = x + y - \left[x + \frac{y(a - x)}{y + (a - x)}\right] = \frac{y^2}{y + (a - x)} > 0.$$

There now remains one last question: which *sign* do we use in the equation for x? After all, the fault is located in one place, of course, not two, and so we can't have two values for x. So, do we use the plus sign, or do we use the minus sign?

Often, when questions like this occur in solving a quadratic equation, we can reject one of the two possibilities on physical grounds (for example, if the solution represents the number of apples in a bag, and the formal solutions are $+3$ and -6, we reject the -6 as physically unreasonable). In the case of a cable fault location, we physically demand that $x > 0$. Alas, as you'll see, *both* signs can result in $x > 0$! So, how *do* we decide which sign to use?

The answer is that we use the minus sign, and here's why. By definition $x < a$ (and so $a - x > 0$), and since the fault resistance must obviously be positive ($y > 0$), these two observations tell us that

$$x = c - \frac{y(a - x)}{y + a - x} < c.$$

Thus, the *unique*, correct solution for x is

$$x = c - \sqrt{(a - c)(b - c)}.$$

More than 30 years ago, when at the Institution of Electrical Engineers in London researching a biography[6] of the English mathematical physicist Oliver Heaviside (1850–1925), I came across one of his notebooks, in which he had recorded (on January 16, 1871) an

---

[6]P. J. Nahin, *Oliver Heaviside*, The Johns Hopkins University Press, 2002. Heaviside, for example, derived the Poynting vector for energy flow in an electromagnetic field before Poynting did (but Poynting published first); it's Heaviside's derivation, however, that's in modern textbooks.

application of this formula. He had made that calculation while working as a telegraph operator in Denmark. Observing that a fault had occurred somewhere along the 360-mile undersea cable between Sondervig in Denmark (the right end of the cable in Figure P26.1) and Newbiggin-by-the-Sea in England (the left end), the two electrical measurements (made at the English end) gave the values $b = 1,040$ ohms and $c = 970$ ohms. Knowing that the cable had a resistance of 6 ohms per mile, Heaviside calculated $a = 6(360) = 2,160$ ohms. Therefore,

$$x = 970 - \sqrt{(2,160 - 970)(1,040 - 970)} = 682 \text{ ohms},$$

or, at 6 ohms per mile, Heaviside concluded that the fault was located $113\frac{2}{3}$ miles from the English end of the cable. At the end of his calculations, Heaviside wrote "All over. Dined roast beef, apple tart, and rabbit pie, with Claret, and enjoyed ourselves." And, on that happy note, you have at last reached the end of this book, or you have once you try your hand at the following challenge problem.

Not all cables are underwater, of course, and in fact buried *pairs* of landline cables are not uncommon. Cables buried in the ground aren't subject to the abrasion caused by being moved around by water currents, or to fish bites. But the protective sheathing of a buried cable can be degraded by corrosion and/or penetrated by underground rodents, and so faults *can* still occur. Suppose we have such a cable pair in a very long trench, perhaps dozens of miles in length, and one of the cables has developed a *short-circuit fault* to ground with some unknown cable resistance x, measured from the common location of the pair's left ends (see Figure P26.2) to the fault.

If we can measure the resistance x in the damaged cable, then we can use the resistance per mile of the cable to compute the location of the spot at which we'll dig up the cable for repair. We can use the fact that we have a good cable, along with the damaged one, to allow a very accurate calculation of x using what is called a *Wheatstone bridge circuit*.[7] Figure P26.2 shows the circuit hookup, where b and c are two

---

[7]Named after the English scientist Charles Wheatstone (1802–1875), although he actually wasn't the inventor (which he freely admitted)! Wheatstone was Oliver Heaviside's uncle by marriage (Wheatstone's wife was the sister of Oliver's mother).

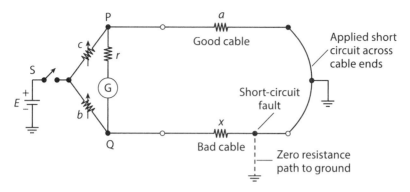

Figure P26.2. Wheatstone bridge circuit for locating a short-circuit fault in one of two cables buried in a common trench.

resistors whose known values can be very accurately adjusted over a wide range of values. The symbol G represents a *galvanometer*, a very sensitive current meter that serves as a *bridge* between the two cable paths. The value of r, the current-limiting resistor in series with G, is unimportant (it's there simply to prevent burning out G if, somehow, a voltage source would be accidentally applied directly to G, which has a very small internal resistance—a fraction of an ohm). The undamaged cable is known to have a total resistance of a ohms.

The circuit is connected as follows. A battery of any voltage (its value E is unimportant) is connected through the switch S to b and c, and b and c are in turn connected to the left ends of the two buried cables. The right ends of the two cables are shorted together, and the short is then connected to an Earth ground. Thus, the right ends of the cables are both electrically connected to the cable fault. When the switch S is up (as shown in Figure P26.2), the battery is not connected to anything, and so G will indicate there is zero current in it.

We then alternately close and open S, and adjust the variable resistors b and c so that the needle of G *always shows zero*. Thus, we are looking for the circuit condition such that as S opens and closes, there is *never* any current in G (this is why the value of r is unimportant, because there is never any current in it, either), either when S is up (open) and so there is no current *anywhere*, or when S is down (closed) and the battery current flows *only* into the cables but not through G. This is a very sensitive test, as we are not attempting to accurately read a precise value of current but, rather, we are searching for those values of b and c that result in not even a *quiver* in G's needle as we flip S up and down. Show how to use the resulting values of b and c to compute x.

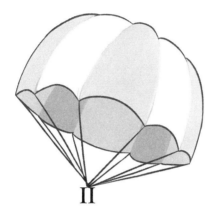

II

THE SOLUTIONS

Challenge Problem 1

No matter what x and y are, $72x$ and $694y$ are each even (because 72 and 694 are even), and the sum of two even numbers is even. But 1,001,001 is odd, and so there can be *no* integer solutions to the first equation. That's it! For the second equation, we start by observing that $x^3 = 4z^3 - 2y^3 = 2(2z^3 - y^3)$, and so, clearly, x^3 is even. But that means x is even (if x were odd, then there is an integer n such that $x = 2n + 1$, and so x^3 would be odd instead of even, so x can't be odd; that is, x must be even). Thus, there is an integer x_1 such that $x = 2x_1$. So, putting that into the original equation we get $8x_1^3 + 2y^3 = 4z^3$, or $y^3 = 2z^3 - 4x_1^3 = 2(z^3 - 2x_1^3)$, and so, clearly, y^3 is even. But that means y is even, and so there is an integer y_1 such that $y = 2y_1$. So, putting that into $8x_1^3 + 2y^3 = 4z^3$ we get $8x_1^3 + 16y_1^3 = 4z^3$, or $z^3 = 2x_1^3 + 4y_1^3 = 2(x_1^3 + 2y_1^3)$, and so, clearly, z^3 is even. But that means z is even and that means there is an integer z_1 such that $z = 2z_1$. So, putting this into $8x_1^3 + 16y_1^3 = 4z^3$ we get $8x_1^3 + 16y_1^3 = 32z_1^3$, and so $x_1^3 + 2y_1^3 = 4z_1^3$. This is the *same equation* we started with (!), which means that if (x,y,z) is a solution, then so is $(x/2, y/2, z/2)$, and so on, *forever*. But there can't be a *forever*, as sooner or later we'll run out of continually decreasing positive integers. So (x,y,z) couldn't have existed in the first place, and so, again, there are *no* integer solutions. This is a particular example of deriving an absurdity by the method of *infinite descent*, a method made famous by the French mathematician Pierre de Fermat (1601–1665) in his proof that there are no integer solutions to $x^4 + y^4 = z^4$ (a special case of the celebrated Fermat's last theorem).

Challenge Problem 2

We write $(1+x)^{1/2} = c_0 + c_1 x + c_2 x^2 + c_3 x^3 + \ldots$ and notice that $(1+x)^{1/2}(1+x)^{1/2} = 1 + x$. Thus, $(c_0 + c_1 x + c_2 x^2 + c_3 x^3 + \ldots)(c_0 + c_1 x + c_2 x^2 + c_3 x^3 + \ldots) = 1 + x$. If we perform the multiplication on the left-hand side of the equality, we get $c_0^2 + 2c_0 c_1 x + (c_1^2 + 2c_0 c_2)x^2 + 2(c_0 c_3 + c_1 c_2)x^3 + (c_2^2 + 2c_1 c_3 + 2c_0 c_4)x^4 + \ldots = 1 + x$. So, equating coefficients of equal powers of x on each side of the equality, $c_0^2 = 1$, or $c_0 = 1$. Also, $2c_0c_1 = 2c_1 = 1$, or $c_1 = \frac{1}{2}$. Also, $(\frac{1}{2})^2 + 2(1)c_2 = 0$, or $c_2 = -\frac{1}{8}$. Also, $2(c_3 - \frac{1}{16}) = 0$, or $c_3 = \frac{1}{16}$. In the same way, you should be able to show that $c_4 = -\frac{5}{128}$, and so on.

PREFACE

Challenge Problem 3

The mass on the incline has weight $2mg$, with a component of that force *normal* to the incline of $2mg \cos(\theta)$, and a component directed *down* the incline of $2mg \sin(\theta)$. Thus, if that mass moves on the incline, it will experience a frictional drag force of $2mg\mu \cos(\theta)$ directed *opposite* its direction of motion. Finally, since we are told that the masses move at a constant speed, they are not accelerating, and so by Newton's second law of motion the net force on each mass is zero. Since the hanging mass has a downward force acting on it (gravity) of mg, then the tension (directed upward) in the string must be mg. This same tension acts on the $2m$ mass as well, directed *up* the incline. Now, suppose the $2m$ mass slides *up* the incline (and so the hanging mass falls). We can then write, for the mass on the incline, $2mg \sin(\theta) + 2mg\mu \cos(\theta) = mg$, where the two terms on the left are directed *down* the incline, and the term on the right is directed *up* the incline. Thus, $\sin(\theta) + \mu \cos(\theta) = \frac{1}{2}$, or $\mu = \frac{\frac{1}{2} - \sin(\theta)}{\cos(\theta)}$. The physically interesting interval for θ is, of course, $0° \leq \theta < 90°$. For $0° \leq \theta \leq 30°$ we have $\mu \geq 0$, which is physically okay. For $30° < \theta < 90°$, however, we have $\mu < 0$, which is *not* physically okay. Next, suppose the $2m$ mass slides *down* the incline (and so the hanging mass rises). We can then write, for the mass on the incline, $2mg \sin(\theta) = 2mg\mu \cos(\theta) + mg$, where the term on the left is directed *down* the incline, and the two terms on the right are directed *up* the incline. Thus, $\sin(\theta) = \mu \cos(\theta) + \frac{1}{2}$, or $\mu = \frac{\sin(\theta) - \frac{1}{2}}{\cos(\theta)}$. For $30° \leq \theta < 90°$ we have $\mu \geq 0$, which is physically okay, while for $0° \leq \theta < 30°$ we have $\mu < 0$, which is *not* physically okay. In Figure 1, the lower curve is θ versus μ for the $2m$ mass sliding *up* the incline, and the upper curve is θ versus μ for the $2m$ mass sliding *down* the incline. For $0 \leq \mu \leq \frac{1}{2}$ the θ, μ relationship is *double valued*; that is, the

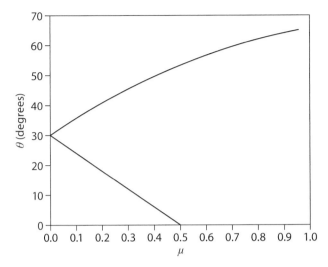

Figure 1. Preface — Sliding on an incline.

$2m$ mass can slide either up *or* down. For $\mu > \frac{1}{2}$, however, the $2m$ mass, *if sliding at a constant speed*, can do so only by sliding *down* the incline (thereby dragging the hanging mass upward). I don't think any of this is obvious by inspection, and so this is an elementary example of the power of *analytical* reasoning.

PREFACE
Challenge Problem 4

Let there be N eggs in the basket *and then add an egg* (remembering the "imaginary cow" trick), giving $N + 1$ eggs in the basket. *Now*, when we draw 2, 3, 4, 5, or 6 eggs at a time we'll empty the basket (before, we were always one egg *short*). So, $N + 1$ is exactly divisible by 2, 3, 4, 5, and 6, while N itself is exactly divisible by 7. We can find $N + 1$ by reasoning as follows. For a numerator to be exactly divisible by a denominator, *all* the factors of the denominator must be present in the factors of the numerator. So, factor 2, 3, 4, 5, and 6 into their prime factors (which means they can't be factored further). That is, $2 = 2$, $3 = 3$, $4 = (2)(2)$, $5 = 5$, and $6 = (2)(3)$. Then, we multiply all the distinct primes by the *maximum* number of times each occurs. In this case, we have $[(2)(2)][3][5] = 60$. This tells us that the numerator $N+1$ must be a multiple of 60. That is, $N+1 = 60$ or 120 or 180 or ... and so on. It's now an easy task to check each of these possible values to see if one less (that is, N) is divisible by 7. We quickly see that 60 doesn't work, but 120 does, as $119/7 = 17$. So, the smallest possible answer is that there are 119 eggs in the basket (before we add our extra egg). For the *next* smallest number of eggs that works, this same approach just as quickly gives us that answer. So, continuing to check multiples of 60 to see if one less than the number is divisible by 7, we find that 540 works ($539/7 = 77$), and so, after 119 eggs, the next possible number of eggs is 539. Doing this one more time we find that 960 works ($959/7 = 13$), and so the next (third) possible number of eggs is 959.

PREFACE
Challenge Problem 5

Now let's first *remove* an egg (the egg that always remains when we remove 2, 3, 4, 5, or 6 at a time) from the N eggs in the basket, which means that the remaining $N-1$ eggs are evenly divisible by 2, 3, 4, 5, or 6 (that is, by the previous "eggs in a basket" solution, $N-1$ now is a multiple of 60), while N itself is evenly divisible by 7. So, again testing the multiples of 60 (60, 120, 180, 240, 300, ...) for these properties, we find that $N-1 = 300$ is the first one that works ($301/7 = 43$). So, there could be 301 eggs in the basket. The next smallest possible value is 721 eggs (720 is a multiple of 60, and $721/7 = 103$), and the next (third) smallest possible value is 1,141 eggs (1,040 is a multiple of 60, and $1,141/7 = 163$).

PREFACE
Challenge Problem 6

Can a line pass through zero lattice points? Yes. Consider $y = x + \sqrt{2}$. There is no integer x that when added to the irrational $\sqrt{2}$ gives an integer y. Can a line pass through *exactly one* lattice point? Yes. Consider $y = x\sqrt{2}$. It passes through $(0,0)$ but not any other lattice

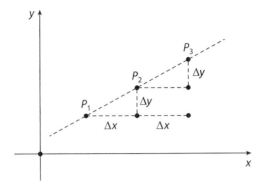

Figure 2. Preface — Drawing a line through the lattice points.

point, as there is no integer $x \neq 0$ that when multiplied by the irrational $\sqrt{2}$ gives an integer $y \neq 0$. Can a line pass through *exactly two* lattice points? No. That's because if a line passes through *two* lattice points it must then pass through an infinity more of lattice points. To see this, simply draw the sketch shown in Figure 2 of the Preface.

That is, if P_1 and P_2 are two given lattice points, displaced by Δx and Δy (which are, of course, both integer quantities), then the line must also pass through the lattice point P_3, which is displaced from P_2 by Δx and Δy. That is, the line segment joining P_2 and P_3 is a *continuation* of the line segment joining P_1 and P_2, and both line segments have the *same slope*, and so on, to points (not shown) P_4 , P_5, ..., forever. The same argument obviously works in the opposite direction as well.

PREFACE
Challenge Problem 7

At the instant the ball starts upward on the Moon, its energy is all kinetic energy (KE), with zero potential energy (PE).[1] As it rises, the ball's KE decreases and its PE increases (with the total remaining constant because of the conservation of energy and because there is no

---

[1] I'm taking the location of your palm as the zero PE reference, for convenience. The zero reference could, of course, be anywhere, and the argument would be unchanged.

energy dissipation mechanism present, like air drag). When the ball reaches its maximum height its energy is all PE with zero KE. The PE at maximum height is thus equal to its initial KE. As the ball falls, its PE at any given point is the same as it was at that point during its ascent, and so its KE is the same, as well (because the total energy is constant). That is, the ball, at every point, moves at the same speed coming down as it did when going up. Thus, as the distance up equals the distance down, the up and down times are equal, too, *on the Moon*. On the Earth, however, things must be modified to account for the continual energy loss due to air drag. In particular, the PE at maximum height will be *less* than the initial KE. And since the PE at any given point in the descent equals what it was during the ascent, the KE at any point during the descent must be *less* than its KE was at that point during the ascent. That is, the ball is always, at every point, moving more slowly during its fall than it was during its rise. So, it takes longer to fall than it did to rise ($t_d > t_u$). Again, notice that the details of air drag are *not import*ant here, just that air drag dissipates energy.

PROBLEM 1
A Military Question

The retarding effect of gravity on the projectile is *only* in the vertical direction, so the horizontal speed of the projectile is constant, equal always to its initial horizontal value, which is $V \cos(\theta)$. The vertical (upward) speed is a different story, however, as it *is* affected by gravity. The initial vertical speed of $V \sin(\theta)$ is continually reduced by gt, where t is the time measured from launch (at $t = 0$). That is, the vertical speed at time t is $V \sin(\theta) - gt$. If we say $t = T$ is the time of impact on the wall, then the projectile will have traveled a horizontal distance of $VT \cos(\theta)$, which of course must be D. The vertical distance (that is, h) of the projectile is the integral of its vertical speed, and so we have the two equations

$$VT \cos(\theta) = D \tag{1}$$

and

$$VT \sin(\theta) - \frac{1}{2}gT^2 = h. \tag{2}$$

Solving (1) for T,

$$T = \frac{D}{V\cos(\theta)}, \tag{3}$$

and then putting this T into (2), we have

$$D\tan(\theta) - \frac{gD^2}{2V^2\cos^2(\theta)} = h. \tag{4}$$

Remembering the trigonometric identity

$$\sin^2(\theta) + \cos^2(\theta) = 1$$

and then dividing through by $\cos^2(\theta)$ gives

$$\tan^2(\theta) + 1 = \frac{1}{\cos^2(\theta)}.$$

Using this result for $1/\cos^2(\theta)$ in (4), we have

$$D\tan(\theta) - \frac{gD^2}{2V^2}\left[\tan^2(\theta) + 1\right] = h,$$

or with just a touch of rearrangement,

$$\frac{gD^2}{2V^2}\tan^2(\theta) - D\tan(\theta) + h + \frac{gD^2}{2V^2} = 0. \tag{5}$$

Equation (5) is a quadratic equation in $\tan(\theta)$, and so by the solution formula for the quadratic equation, we can immediately write

$$\tan(\theta) = \frac{D \pm \sqrt{D^2 - 4\frac{gD^2}{2V^2}\left(h + \frac{gD^2}{2V^2}\right)}}{2\frac{gD^2}{2V^2}},$$

or making the obvious simplifications,

$$\tan(\theta) = \frac{V^2}{gD^2}\left[D \pm \sqrt{D^2 - 2\frac{gD^2}{V^2}\left(h + \frac{gD^2}{2V^2}\right)}\right]. \tag{6}$$

Now, this is where we make a *crucial* observation. It is *physically* required that $\tan(\theta)$ be *real*-valued (after all, just how *would* you launch a projectile at a *complex*-valued angle?), which means the expression in (6) under the square-root sign must not be negative. So, it must be true that

$$D^2 - 2\frac{gD^2}{V^2}\left(h + \frac{gD^2}{2V^2}\right) \geq 0,$$

which is easily rearranged to give

$$h \leq \frac{V^2}{2g}\left[1 - \left(\frac{gD}{V^2}\right)^2\right]. \tag{7}$$

Equation (7) immediately tells us that h has its maximum value when the inequality is an equality, and so writing the maximum h as H, we have

$$\boxed{H = \frac{V^2}{2g}\left[1 - \left(\frac{gD}{V^2}\right)^2\right].}$$

For this maximum value of h, the square root in (6) is zero, and so we next have the required launch angle to be $\theta = \theta_{max}$, where

$$\boxed{\theta_{max} = \tan^{-1}\left(\frac{V^2}{gD}\right).}$$

Now, what about the flight time, T? We've just about got that, too. From (3) we'd have T *immediately* if we had $\cos(\theta_{max})$. This is easy to determine from our last boxed equation by simply drawing the right triangle in Figure S1.1 and using the Pythagorean theorem to get the hypotenuse.

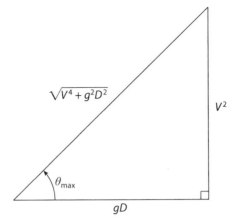

Figure S1.1. Determining $\cos(\theta_{\max})$.

From Figure S1.1 we have

$$\cos(\theta_{\max}) = \frac{gD}{\sqrt{V^4 + g^2 D^2}}$$

and substituting this into (3) gives us our final answer:

$$T = \sqrt{\left(\frac{V}{g}\right)^2 + \left(\frac{D}{V}\right)^2}.$$

Just to give you a feel for the numbers, suppose $D = 500$ feet, $V = 200$ feet per second, and $g = 32$ feet per second-squared. Then,

$$\theta_{\max} = 68.2°,$$
$$H = 525 \text{ feet},$$
$$T = 6.73 \text{ seconds}.$$

PROBLEM 2
A Seemingly Impossible Question

To be able to translate the snowplow problem into mathematics, I'll make the following plausible *physical* assumptions. If we measure time from noon ($t = 0$ *is* noon), then the snow started falling at some earlier time $t = -T$, where $T > 0$. Clearly, the determination of T is our goal. To that end, let's denote the distance traveled by the plow at time t by $y(t)$, and so $\frac{dy}{dt}$ is the speed of the plow, as well as $y(0) = 0$. To be specific about units (we are, of course, free to choose whatever units we wish), let's measure y in miles and t in hours (and so $\frac{dy}{dt}$ has the units of miles per hour). Let's further measure the depth of the snow at time t by $x(t)$, and so (using consistent units) the depth is in miles, admittedly a rather nonstandard unit for snow depth on a city street, but there is nothing wrong with doing that. Next, let's denote the width of the snowplow blade by W (in units, again, of miles), and the plow's fixed snow-removal rate by K (cubic miles per hour). For the next step, I'll make the assumption (plausible, I think) that the height of the blade is always greater than the snow depth. This last assumption allows us to write our first equation:

$$Wx\frac{dy}{dt} = K, \tag{1}$$

which you'll notice has the same units (cubic miles per hour) on both sides of the equation. For a final assumption, let's take the snowfall rate as S miles per hour. Since at time $t \geq 0$ the snow has been falling for a time interval of $t + T$, we can now write our second equation:

$$x = S(t + T). \tag{2}$$

Inserting (2) into (1), we get

$$WS(t + T)\frac{dy}{dt} = K,$$

or

$$\frac{dy}{dt} = \frac{K}{WS}\frac{1}{t + T}.$$

Since $\frac{K}{WS}$ is just a combination of constants and so is, itself, a constant, let's replace it with a single constant A, and thus

$$\frac{dy}{dt} = \frac{A}{t+T},$$

or if we separate variables—that is, get y alone on one side of the equality and t alone on the other side—we have

$$dy = A\frac{dt}{t+T}. \tag{3}$$

We can integrate (3) indefinitely (no integration limits) to get

$$y = A \int \frac{dt}{t+T} + C, \tag{4}$$

where C is the constant of indefinite integration (differentiating (4) with respect to t gives (3) back for *any* C). We'll pick the "right" C in just a moment. To evaluate the integral on the right in (4), let's change variables to $u = t+T$ (and so $du = dt$), which transforms (4) to

$$y = A \int \frac{du}{u} + C = A \ln(u) + C. \tag{5}$$

Now, there must be some constant B such that $C = A \ln(B)$, and so

$$y(t) = A \ln(t+T) + A \ln(B) = A\left[\ln(t+T) + \ln(B)\right],$$

or by operations of logarithms

$$y(t) = A \ln\{B(t+T)\}. \tag{6}$$

Well, what *are* A and B? First, we know $y = 0$ at $t = 0$, and so putting this so-called initial condition into (6), we have

$$0 = A \ln(BT),$$

or since $A \neq 0$ (look back at its definition),

$$\ln(BT) = 0.$$

which means $BT = 1$. That is, $B = 1/T$, and so

$$y(t) = A \ln \left\{ \frac{1}{T}(t + T) \right\} = A \ln \left\{ 1 + \frac{t}{T} \right\}. \tag{7}$$

To get A, we use the given information about where the plow is at $t = 1$ hour and at $t = 2$ hours. If we write D as the distance (in miles) for $y(1)$, then $y(2) = \frac{3}{2} D$. So, putting these two *boundary conditions* into (7), we have

$$D = A \ln \left\{ 1 + \frac{1}{T} \right\}, \tag{8}$$

and

$$\frac{3}{2} D = A \ln \left\{ 1 + \frac{2}{T} \right\}. \tag{9}$$

Dividing (9) by (8), we see both D and A cancel, leaving us (mercifully!) with a single equation for T:

$$\frac{3}{2} = \frac{\ln \left\{ 1 + \frac{2}{T} \right\}}{\ln \left\{ 1 + \frac{1}{T} \right\}},$$

or

$$3 \ln \left\{ 1 + \frac{1}{T} \right\} = 2 \ln \left\{ 1 + \frac{2}{T} \right\},$$

or

$$\ln \left\{ 1 + \frac{1}{T} \right\}^3 = \ln \left\{ 1 + \frac{2}{T} \right\}^2,$$

and so

$$\left\{ 1 + \frac{1}{T} \right\}^3 = \left\{ 1 + \frac{2}{T} \right\}^2.$$

This means

$$\frac{(T + 1)^3}{T^3} = \frac{(T + 2)^2}{T^2},$$

or

$$(T + 1)^3 = T(T + 2)^2.$$

Multiplying out both sides, and making the obvious algebraic simplifications, we arrive at a quadratic equation for T:

$$T^2 + T - 1 = 0,$$

which we know how to solve:

$$T = \frac{-1 \pm \sqrt{1+4}}{2} = \frac{-1 \pm \sqrt{5}}{2} \text{ hours,}$$

or using the plus sign (because it gives $T > 0$, while the negative sign gives $T < 0$ which is not *physically* correct) we arrive at $T = 0.618033\ldots$ hours. That is, $T = 37$ minutes, 5 seconds, and since it started to snow this much time before noon, the answer to our question is,

> it started to snow at $11:22:55$ am.

PROBLEM 3
Two Math Problems

(1): The key observation is that since a, b, and c are the sides of a triangle, they must satisfy the so-called triangle inequalities, the mathematical statement of the physical fact that the sum of any two sides is equal to or greater than the remaining side. That is, the *shortest* path between any two points (any two of the vertices of a triangle) in a plane is a straight line (the side joining the two vertices). So, we have, for the sides of any triangle,

$$a + b \geq c, \tag{1}$$

and

$$b + c \geq a, \tag{2}$$

and

$$a + c \geq b. \tag{3}$$

From (1) we have $a \geq c - b$, and from (3) we have $a \geq b - c$. Now, since a, b, and c are, physically, *positive* numbers, then either $c - b \geq 0$, and so $b - c \leq 0$, or vice versa. Whichever is the nonnegative value, a is greater than or equal to it and so is certainly greater than or equal to the nonpositive value. We can therefore write

$$\sqrt{\{a + (b - c)\}\{a - (b - c)\}} = \sqrt{a^2 - (b - c)^2} \leq a \tag{4}$$

and be assured that we are always taking the square root of a nonnegative number. From (1) we have $b \geq c - a$, and from (2) we have $b \geq a - c$. Now, either $c - a \geq 0$, and so $a - c \leq 0$, or vice versa. Whichever is the nonnegative value, b is greater than or equal to it and so is certainly greater than or equal to the nonpositive value. We can therefore write

$$\sqrt{\{b + (c - a)\}\{b - (c - a)\}} = \sqrt{b^2 - (c - a)^2} \leq b \tag{5}$$

and be assured that we are always taking the square root of a nonnegative number. And finally, from (2) we have $c \geq a - b$, and from (3) we have $c \geq b - a$. Now, either $a - b \geq 0$, and so $b - a \leq 0$, or vice versa. Whichever is the nonnegative value, b is greater than or equal to it and so is certainly greater than or equal to the nonpositive value. We can therefore write

$$\sqrt{\{c + (a - b)\}\{c - (a - b)\}} = \sqrt{c^2 - (a - b)^2} \leq c \tag{6}$$

and be assured that we are always taking the square root of a nonnegative number. Multiplying (4), (5), and (6) together gives

$$\sqrt{\{a+(b-c)\}\{a-(b-c)\}\{b+(c-a)\}\{b-(c-a)\}\{c+(a-b)\}\{c-(a-b)\}}$$
$$\leq abc,$$

or

$$\sqrt{(a+b-c)(a-b+c)(b+c-a)(b-c+a)(c+a-b)(c-a+b)}$$
$$= \sqrt{(a+b-c)^2(b+c-a)^2(c+a-b)^2} \leq abc,$$

and so

$$abc \geq (a+b-c)(b+c-a)(c+a-b),$$

and we are done. This result is called *Padoa's inequality*, after the Italian mathematician Alessandro Padoa (1868–1937).

(2): In $I = \int_0^1 x \left\{\frac{1}{x}\right\} dx$ change the variable to $y = \frac{1}{x}$. Then, $\frac{dy}{dx} = -\frac{1}{x^2}$ and so $dx = -x^2 dy = -\frac{dy}{y^2}$. Thus, since y varies from ∞ to 1 as x varies from 0 to 1, we have

$$I = \int_\infty^1 \frac{1}{y} \{y\} \left(-\frac{dy}{y^2}\right)$$

$$= \int_1^\infty \frac{\{y\}}{y^3} dy.$$

The "trick" is now to write the integral as an infinite sum of integrals, each integral over a unit interval. That is,

$$I = \sum_{k=1}^\infty \int_k^{k+1} \frac{\{y\}}{y^3} dy,$$

or using $\{y\} = y - [y]$, we have

$$I = \sum_{k=1}^\infty \int_k^{k+1} \frac{y - [y]}{y^3} dy.$$

The reason for doing this is that in each integral as y varies from k to $k + 1$, $[y] = k$, a *constant*, over the *entire* unit interval of integration. So,

$$I = \sum_{k=1}^\infty \int_k^{k+1} \frac{y - k}{y^3} dy = \sum_{k=1}^\infty \left(\int_k^{k+1} \frac{dy}{y^2} - k \int_k^{k+1} \frac{dy}{y^3}\right),$$

where each of the two integrals on the far right is easy to do. Specifically,

$$\int_k^{k+1} \frac{dy}{y^2} = \left(-\frac{1}{y}\right)\Big|_k^{k+1} = \frac{1}{k} - \frac{1}{k+1},$$

and

$$k \int_k^{k+1} \frac{dy}{y^3} = k \left(-\frac{1}{2y^2} \right) \Big|_k^{k+1} = \frac{k}{2} \left(\frac{1}{k^2} - \frac{1}{(k+1)^2} \right) = \frac{1}{2} \left(\frac{1}{k} - \frac{k}{(k+1)^2} \right).$$

Thus,

$$I = \sum_{k=1}^{\infty} \left(\frac{1}{k} - \frac{1}{k+1} \right) - \frac{1}{2} \sum_{k=1}^{\infty} \left(\frac{1}{k} - \frac{k}{(k+1)^2} \right).$$

The first sum on the right presents little difficulty. We just write it out, term by term, to get

$$\sum_{k=1}^{\infty} \left(\frac{1}{k} - \frac{1}{k+1} \right) = \left(1 - \frac{1}{2} \right) + \left(\frac{1}{2} - \frac{1}{3} \right) + \left(\frac{1}{3} - \frac{1}{4} \right) + \cdots ,$$

and we see that all the terms self-cancel, *except* for the very first one, leaving just the starting 1. Thus,

$$I = 1 - \frac{1}{2} \sum_{k=1}^{\infty} \left(\frac{1}{k} - \frac{k}{(k+1)^2} \right).$$

To do this final sum is almost as easy, once we see that

$$\sum_{k=1}^{\infty} \left(\frac{1}{k} - \frac{k}{(k+1)^2} \right) = \sum_{k=1}^{\infty} \left(\frac{1}{k} - \frac{k+1}{(k+1)^2} + \frac{1}{(k+1)^2} \right)$$

$$= \sum_{k=1}^{\infty} \left(\frac{1}{k} - \frac{1}{k+1} + \frac{1}{(k+1)^2} \right)$$

$$= \sum_{k=1}^{\infty} \left(\frac{1}{k} - \frac{1}{k+1} \right) + \sum_{k=1}^{\infty} \frac{1}{(k+1)^2},$$

or recalling the earlier result,

$$\sum_{k=1}^{\infty} \left(\frac{1}{k} - \frac{1}{k+1} \right) = 1,$$

we have

$$I = 1 - \frac{1}{2}\left(1 + \sum_{k=1}^{\infty} \frac{1}{(k+1)^2}\right) = 1 - \frac{1}{2}\sum_{k=1}^{\infty}\frac{1}{k^2},$$

and recalling Euler's sum, we see that

$$\boxed{\int_0^1 x\left\{\frac{1}{x}\right\}dx = 1 - \frac{\pi^2}{12},}$$

which has the numerical value I gave you in the original problem statement. If you make the obvious (slight) adjustments to this analysis to calculate $\int_0^1 x^n \left\{\frac{1}{x}\right\}dx$, you'll find the result is

$$\int_0^1 x^n\left\{\frac{1}{x}\right\}dx = \frac{1}{n} - \frac{\zeta(n+1)}{n+1},$$

which, for $n = 1$, reduces to the result in the box. For $n = 2$ the result involves $\zeta(3)$, for which there is no known exact result, but for $n = 3$, Euler showed $\zeta(4) = \frac{\pi^4}{90}$, and so

$$\int_0^1 x^3\left\{\frac{1}{x}\right\}dx = \frac{1}{3} - \frac{\zeta(4)}{4} = \frac{1}{3} - \frac{\pi^4}{360} = 0.06275\ldots.$$

And so now you know how mathematical physicists pass the time when a really boring commercial appears on the TV screen!

PROBLEM 4
An Escape Problem

Let $t = 0$ be the time you decide to start your escape from the truck by running at angle θ to the center of the street. Then, at time $t = T > 0$, given by $T = \frac{S}{V_t - V_y \cos(\theta)}$, the truck passes you (because your speed component *along* the street is $V_y \cos(\theta)$). During the time interval $0 \leq t \leq T$, you will have moved away from the center of the street by distance

$$TV_y \sin(\theta) = \frac{SV_y \sin(\theta)}{V_t - V_y \cos(\theta)},$$

because your speed component *directly away* from the center of the street is $V_y \sin(\theta)$. That is, denoting this distance by y, we have

$$y = \frac{SV_y \sin(\theta)}{V_t - V_y \cos(\theta)},$$

and we wish to pick θ to maximize y. At this point, most people who have had first-year calculus think "let's set the derivative of y with respect to θ to zero." That *will* work, too, but in fact we can find the optimal θ *without* calculus, using just algebra and the quadratic formula. Squaring the y-equation, and collecting terms (remembering, too, that $\cos^2(\theta) + \sin^2(\theta) = 1$), we arrive at a quadratic in $\cos(\theta)$:

$$V_y^2 \left[y^2 + S^2\right] \cos^2(\theta) - 2y^2 V_t V_y \cos(\theta) + \left(y^2 V_t^2 - S^2 V_y^2\right) = 0.$$

From the quadratic formula, we have

$$\cos(\theta) = \frac{2y^2 V_t V_y \pm \sqrt{4y^4 V_t^2 V_y^2 - 4V_y^2 \left[y^2 + S^2\right]\left[y^2 V_t^2 - S^2 V_y^2\right]}}{2V_y^2 \left[y^2 + S^2\right]},$$

which *must* involve the square root of a *nonnegative* quantity to give a *physically reasonable* θ (to give a *real* value to $\cos(\theta)$). So, setting

$$4y^4 V_t^2 V_y^2 - 4V_y^2 \left[y^2 + S^2\right]\left[y^2 V_t^2 - S^2 V_y^2\right] \geq 0,$$

we get

$$y^4 V_t^2 \geq \left[y^2 + S^2\right]\left[y^2 V_t^2 - S^2 V_y^2\right] = y^4 V_t^2 - S^2 y^2 V_y^2 + S^2 y^2 V_t^2 - S^4 V_y^2,$$

or

$$0 \geq y^2 \left(V_t^2 - V_y^2 \right) - S^2 V_y^2,$$

or

$$\frac{S^2 V_y^2}{V_t^2 - V_y^2} \geq y^2,$$

or

$$y \leq \frac{SV_y}{\sqrt{V_t^2 - V_y^2}}.$$

So, the *maximum* distance from the center of the street is

$$\frac{SV_y}{\sqrt{V_t^2 - V_y^2}}.$$

Putting this y into the result in the box for $\cos(\theta)$, we know the square root will be zero, and so the optimal escape angle is

$$\cos \left(\theta_{opt} \right) = \frac{2Y^2 V_t V_y}{2V_y^2 \left[Y^2 + S^2 \right]} = \frac{\frac{S^2 V_y^2}{V_t^2 - V_y^2} V_t}{V_y \left[\frac{S^2 V_y^2}{V_t^2 - V_y^2} + S^2 \right]} = \frac{V_y}{V_t},$$

or

$$\theta_{opt} = \cos^{-1} \left(\frac{V_y}{V_t} \right).$$

So, for the truck going 60 miles per hour (88 feet per second), and a jogger running at 15 miles per hour (22 feet per second), we have

$$\theta_{opt} = \cos^{-1} \left(\frac{22}{88} \right) = \cos^{-1} \left(\frac{1}{4} \right) = 75.5°.$$

For the truck 75 feet behind the jogger when the escape starts, $Y = \frac{75(22)}{\sqrt{88^2 - 22^2}}$ feet $= 19.4$ feet > 4 feet (half the width of the truck, and so the jogger survives. For a jogger running at just 4 miles per hour (5.87 feet per second), $\theta_{opt} = 86.2°$, and $Y = 5$ feet, which, while again equaling survival, will probably be remembered as a near-death experience.

PROBLEM 5
The Catapult Again

A projectile launched from $x = 0$ at time $t = 0$, at angle θ, with initial speed V, has a *constant* horizontal speed component of $V \cos(\theta)$ and a vertical speed component of $V \sin(\theta) - gt$. Thus, the position of the projectile for any $t \geq 0$, until it returns to Earth, is given by

$$x(t) = V \cos(\theta)t \ \text{ and } \ y(t) = V \sin(\theta)t - \frac{1}{2}gt^2.$$

When the projectile returns to Earth at time $t = T$ we'll of course have $y(T) = 0$, and so

$$V \sin(\theta)T - \frac{1}{2}gT^2 = 0,$$

or

$$T = \frac{2V \sin(\theta)}{g}.$$

The horizontal distance traveled by the projectile from launch until it returns to Earth (that is, its *range*) is

$$x(t = T) = V \cos(\theta)\frac{2V \sin(\theta)}{g} = \frac{2V^2}{g} \cos(\theta)\sin(\theta) = \frac{V^2}{g}\sin(2\theta),$$

which is obviously maximum at $\theta = 45°$, and so the *maximum possible range*, which I'll call R, is

$$R = \frac{2V^2}{g}\left(\frac{1}{\sqrt{2}}\right)\left(\frac{1}{\sqrt{2}}\right) = \frac{V^2}{g}.$$

Thus, $V^2 = Rg$. Now, since $= \frac{x}{V\cos(\theta)}$, we have

$$y(t) = V \sin(\theta)\frac{x}{V \cos(\theta)} - \frac{gx^2}{2V^2 \cos^2(\theta)} = x\tan(\theta) - \frac{gx^2}{2V^2}\left[\frac{1}{\cos^2(\theta)}\right].$$

Since $\sin^2(\theta) + \cos^2(\theta) = 1$, then

$$\frac{\sin^2(\theta)}{\cos^2(\theta)} + 1 = \frac{1}{\cos^2(\theta)} = 1 + \tan^2(\theta).$$

Thus,

$$y = x \tan(\theta) - \frac{gx^2}{2V^2}\left[1 + \tan^2(\theta)\right],$$

or, as $V^2 = Rg$,

$$y = x \tan(\theta) - \frac{x^2}{2R}\left[1 + \tan^2(\theta)\right].$$

Now, let's bring the wall, of height H at $x = D$, into the picture. For the projectile to land on the opposite side of the wall it must (obviously!) *clear the top of the wall*. That is, we must have $y(x = D) \geq H$. So,

$$D\tan(\theta) - \frac{D^2}{2R}\left[1 + \tan^2(\theta)\right] \geq H,$$

or

$$\tan(\theta) - \frac{1}{2}\left(\frac{D}{R}\right)\left[1 + \tan^2(\theta)\right] \geq \frac{H}{D}.$$

Clearly, θ must be greater than zero for this inequality to be true (the right-hand side is always positive, while the left-hand side is negative at $\theta = 0$). If θ is too large, however, the inequality also fails for the same reason. So, there is some *interval* for θ over which the inequality holds. When the inequality *just* holds (that is, it is an *equality*), we have a quadratic in $\tan(\theta)$:

$$\tan(\theta) - \frac{1}{2}\left(\frac{D}{R}\right) - \frac{1}{2}\left(\frac{D}{R}\right)\tan^2(\theta) = \frac{H}{D},$$

or

$$\frac{1}{2}\left(\frac{D}{R}\right)\tan^2(\theta) - \tan(\theta) + \frac{1}{2}\left(\frac{D}{R}\right) + \frac{H}{D} = 0.$$

Figure S5.1. Two ways to *just* clear the wall.

From the quadratic formula (I'll let you verify the easy algebra) we get

$$\theta = \tan^{-1}\left\{\frac{1 \pm \sqrt{1 - \left(\frac{D}{R}\right)^2 - 2\left(\frac{H}{R}\right)}}{\left(\frac{D}{R}\right)}\right\}.$$

The two solutions, one for each choice of the $\pm$ sign, represent the projectile just clearing the top of the wall, as shown in Figure S5.1. The hash-marked interval on the x-axis shows where it is possible for the projectile to land on the far side of the wall.

Notice the tacit assumption we've made: the wall is not *so* high that it's impossible for the projectile to clear the wall at *some* angle of launch. That is, the angle computed from the boxed equation is real-valued, which means the quantity in the square root is not negative. So, we are assuming that $1 - \left(\frac{D}{R}\right)^2 - 2\left(\frac{H}{R}\right) \geq 0$, which implies that $H \leq \frac{R}{2} - \frac{D^2}{2R}$. To show you how the numbers work, suppose $V = 200$ feet per second, $D = 500$ feet, and $H = 100$ feet. Using our last result, we see that it *is* possible for the projectile to clear the wall, as $R = \frac{V^2}{g} = \frac{40,000}{32} = 1,250$ feet, and so H can be no higher than $\frac{1,250}{2} - \frac{250,000}{2,500} = 625 - 100 = 525$ feet (which is appreciably more than 100 feet). Using the boxed equation, we find the high and low launch angles are, respectively, $77.6°$ and $23.7°$. Finally, the hash-marked interval in Figure S5.1 (where it is physically possible for a projectile to land) is 523 feet $< x < 1,250$ feet. (The lower limit is given by $x_{\min} = \frac{2V^2}{g}\cos(77.6°)\sin(77.6°) = 523$.) In particular, there is a 23-foot-wide zone extending from the right-hand side of the wall into which it is impossible for a projectile to land. I was initially tempted to call this a "dead zone" but, of course, that is just where you should stand to *avoid* risking death.

PROBLEM 6
Another Math Problem

The power series expansion of $e^x = 1 + x + \frac{x^2}{2!} + \frac{x^3}{3!} + \cdots$, and so, for any $x > 0$, every term on the right in the infinite series is positive. Thus, the partial sums continually increase, and so if we truncate the infinite series at any point, we'll always get something *less* than e^x. Thus, in particular, $e^x > 1 + x$. Now, for the first question, let $x = \frac{\pi}{e} - 1$, which is clearly positive, since $\pi > e$ (because $\pi > 3$ and $e < 3$). So, $e^{\frac{\pi}{e}-1} > 1 + \frac{\pi}{e} - 1 = \frac{\pi}{e}$, or $\frac{e^{\frac{\pi}{e}}}{e} > \frac{\pi}{e}$, or $e^{\frac{\pi}{e}} > \pi$. Finally, raising both sides to the e-power, we have our first answer:

$$\boxed{e^\pi > \pi^e.}$$

Checking with a calculator, we find that, indeed,

$$e^\pi = 23.14\ldots > \pi^e = 22.45\ldots.$$

For the second question, let $x = \frac{3}{e} - 1$, which is positive, as $\frac{3}{e} > 1$. Thus, $e^{\frac{3}{e}-1} > 1 + \frac{3}{e} - 1 = \frac{3}{e}$, and so, $\frac{e^{\frac{3}{e}}}{e} > \frac{3}{e}$, or $e^{\frac{3}{e}} > 3$, and so, raising both sides to the e-power, we have our second answer:

$$\boxed{e^3 > 3^e.}$$

Checking with a calculator, we find that, indeed,

$$e^3 = 20.08\ldots > 3^e = 19.81\ldots.$$

To answer the third question (which is larger, 3^π or π^3 ?) requires just a little bit of additional work. We write $3^x = e^{\ln(3^x)} = e^{x\ln(3)}$, and so, since $e^y = 1 + y + \frac{y^2}{2!} + \frac{y^3}{3!} + \cdots$, then if we let $y = x\ln(3)$, we have $3^x = 1 + x\ln(3) + \frac{x^2\ln^2(3)}{2!} + \cdots$. Now, since $3 > 1$, then $\ln(3) > 0$, and so, for any $x > 0$, all the terms on the right in the infinite series are positive. Thus, for any $x > 0$ we can certainly write $3^x > 1 + x\ln(3)$. Since $3 > e$, then $\ln(3) > 1$, and so an even stronger inequality is that for any $x > 0$, $3^x > 1 + x$. Now, let $x = \frac{\pi}{3} - 1 > 0$, as $\pi > 3$. So, $3^{\frac{\pi}{3}-1} > 1 + \frac{\pi}{3} - 1 = \frac{\pi}{3}$, or $\frac{3^{\frac{\pi}{3}}}{3} > \frac{\pi}{3}$, or $3^{\frac{\pi}{3}} > \pi$. Cubing both sides, we have the answer to the third question:

$$\boxed{3^\pi > \pi^3.}$$

Checking with a calculator, we find that, indeed,

$$3^\pi = 31.544\ldots > \pi^3 = 31.006\ldots.$$

To answer the last question (which is larger, e^2 or 2^e ?), we can start as before with $e^x = 1 + x + \frac{x^2}{2!} + \frac{x^3}{3!} + \cdots$, but we *cannot* set $x = \frac{2}{e} - 1$ and argue that all the terms in the infinite series are positive (because $x < 0$, since $e > 2$). What we can do is set $x = \frac{2}{e}$ (which is positive) and use a truncated series *with two extra terms*. That is, $e^x > 1 + x + \frac{x^2}{2} + \frac{x^3}{6}$ for any $x > 0$. So, with $x = \frac{2}{e}$ it is certainly true that $e^{\frac{2}{e}} > 1 + \frac{2}{e} + \frac{4}{2e^2} + \frac{8}{6e^3}$. If we replace each e on the right with anything *greater* than e, then each term after the first will be *smaller*, and the inequality becomes even stronger. We know $e = 2.71\ldots$, and so $< 2.75 = \frac{11}{4}$. Thus, $e^{\frac{2}{e}} > 1 + \frac{8}{11} + \frac{32}{121} + \frac{512}{7,986} = 1 + \frac{5,808+2,112+512}{7,986} = 1 + \frac{8,432}{7,986} > 2$. Thus, $e^{\frac{2}{e}} > 2$, or raising both sides to the e-power,

$$\boxed{e^2 > 2^e.}$$

Checking with a calculator, we find that, indeed,

$$e^2 = 7.389\ldots > 2^e = 6.58\ldots.$$

And finally, here's a proof that e is irrational. We start with

$$e = 1 + 1 + \frac{1}{2!} + \frac{1}{3!} + \frac{1}{4!} + \cdots .$$

Thus, it is *immediately* clear from just the first three terms that $e > 2$. Also, as

$$\frac{1}{2!} = \frac{1}{2}, \quad \frac{1}{3!} = \frac{1}{(2)(3)} < \frac{1}{(2)(2)} = \frac{1}{2^2},$$

$$\frac{1}{4!} = \frac{1}{(2)(3)(4)} < \frac{1}{(2)(2)(2)} = \frac{1}{2^3}, \cdots$$

then $e < 2 + \frac{1}{2} + \frac{1}{2^2} + \frac{1}{2^3} + \cdots$, and since $\frac{1}{2} + \frac{1}{2^2} + \frac{1}{2^3} + \cdots = 1$, then $e < 3$. Thus, $2 < e < 3$, from which we conclude that e is not an integer (there are no integers between 2 and 3). Now, suppose e is rational, that is, we *assume* there are integers p and q such that $e = \frac{p}{q}$. Since e is not an integer, we know that $q \geq 2$ ($q = 1$ would give $e = p$, an integer, which we just showed is not true). Now, we multiply through the infinite series for e by $q!$ to get

$$eq! = q! + q! + \frac{q!}{2!} + \frac{q!}{3!} + \cdots + \frac{q!}{q!} + \frac{q!}{(q+1)!} + \frac{q!}{(q+2)!} + \cdots ,$$

or

$$eq! = \{2q! + (3 \cdot 4 \cdots q) + (4 \cdot 5 \cdots q) + \cdots + 1\}$$

$$+ \left\{ \frac{1}{(q+1)} + \frac{1}{(q+1)(q+2)} + \cdots \right\} .$$

On the left, $eq! = \frac{p}{q}q! = p(q-1)!$, which is clearly an integer (by the *assumption* of e being rational). On the right-hand side, the expression in the first pair of curly brackets is clearly an integer. But what of the expression in the second pair of curly brackets? That is, what can we say of

$$S = \frac{1}{(q+1)} + \frac{1}{(q+1)(q+2)} + \frac{1}{(q+1)(q+2)(q+3)} + \cdots = ?$$

Since $q \geq 2$, then replacing each term on the right with a *larger* term, we have (by summing the geometric series on the right in the following):

$$S < \frac{1}{3} + \frac{1}{3^2} + \frac{1}{3^3} + \cdots = \frac{1}{2}.$$

So, all of this says that $eq!$ (which is an integer) is the sum of an integer and something else which is less than $\frac{1}{2}$. But that is obviously absurd. So, our initial assumption that e is rational must be wrong. Thus, e is irrational.

PROBLEM 7
If Theory Fails

The MATLAB code **target.m** "solves" the "three arrows in a circular target" problem. With no loss of generality, the code assumes the target has a radius of 1. The code begins each of 10 million simulations by finding six random numbers (stored in the vector *arrow*) to represent the x, y coordinates of the three arrows. (The x and y coordinates of the first arrow are $arrow(1)$ and $arrow(2)$, respectively, and so on.) This is done by imagining a circle of radius 1, enclosed by a square of side length 2 (from -1 to 1, in the x- and y-directions) centered on the origin. Then, after generating a pair of numbers, each from a uniform distribution over the interval -1 to 1), the pair is accepted if $x^2 + y^2 <$ 1, which means the point (x, y) is inside the circle. Once the locations of the three arrows are found, the distances between the three possible pairs are computed (actually, the distances *squared*, which avoids doing time-consuming square-root operations). If all three distances (squared) are less than 1 (the radius squared) the code says that's a success. When run multiple times, the code produced estimates for the probability that no two of the arrows are separated by a distance greater than the length of the target's radius of between 0.274053 to 0.274552. I don't have a theoretical solution. If you develop one, please send it to me.

```
%target.m/created by PJNahin for Mathematical Physics
rand('state',100*sum(clock));success=0;
for n=1:10000000
   index=0;
   for loop=1:3
      go=0;
      while go==0
         x=-1+2*rand;y=-1+2*rand;
         if x^2+y^2<1
            index=index+1;
            arrow(index)=x;
            index=index+1;
            arrow(index)=y;
            go=1;
         end
      end
   end
   d1=(arrow(1)-arrow(3))^2+(arrow(2)-arrow(4))^2;
   d2=(arrow(1)-arrow(5))^2+(arrow(2)-arrow(6))^2;
   d3=(arrow(3)-arrow(5))^2+(arrow(4)-arrow(6))^2;
   if d1<1&d2<1&d3<1
      success=success+1;
   end
end
success/10000000
```

The simulation of the "line in the circular quadrant" problem is just a bit more complicated, as you've probably guessed by looking at the code **crossings.m**.

```
%crossings.m/created by PJNahin for Mathematical Physics
rand('state',100*sum(clock));ZERO=0;ONE=0;TWO=0;
for n=1:10000000
```

(continued)

(continued)

```
go=0;
while go==0
   x=rand;y=rand;
   if x^2+y^2<1
      x1=x;y1=y;
      go=1;
   end
end
go=0;
while go==0
   x=rand;y=rand;
   if x^2+y^2<1
      x2=x;y2=y;
      go=1;
   end
end
m=(y2-y1)/(x2-x1);
if m>0
   ONE=ONE+1;
else
   b=y1-m*x1;
   den=1+m^2;
   num1=-m*b;root=sqrt(m^2+1-b^2);
   x1=(num1+root)/den;x2=(num1-root)/den;
   y1=m*x1+b;y2=m*x2+b;
   px=x1*x2;py=y1*y2;
   if px<0&py<0
      ZERO=ZERO+1;
   elseif px<0&py>0
      ONE=ONE+1;
   elseif px>0&py<0
      ONE=ONE+1;
   else
```

(continued)

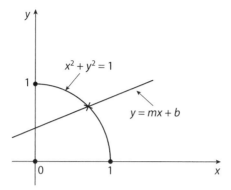

Figure S7.1. Positive slope *always* means *one* intersection.

```
                                        (continued)
        TWO=TWO+1;
    end
  end
end
ZERO/10000000
ONE/10000000
TWO/10000000
```

The code assumes both that the quadrant is the first, and the circle has radius 1, with no loss of generality by either assumption. The code starts by finding two random points in the circular quadrant, by finding two random points in the square $0 < x < 1$, $0 < y < 1$, and accepting them only if $x^2 + y^2 < 1$. These are the two points through which a line will be drawn. Now, before doing anything more, you should immediately see from Figure S7.1 that if the line has positive slope, (more precisely, nonnegative slope), then there will be, *always*, exactly *one* intersection of the line with the circular arc, and so no further calculations are required.

If the slope is negative, however, then the situation is more involved. Since the equation of the full circle is $x^2 + y^2 = 1$, and that of the line is $y = mx + b$, where m is the slope, then the two intersection points of

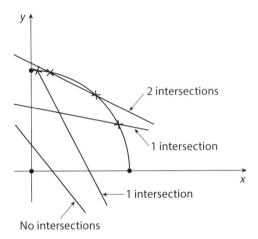

Figure S7.2. The four possibilities for lines with negative slope and their intersections with the first-quadrant circular arc.

the line with the *full circle* have x-coordinates given by the solutions to $x^2 + (mx + b)^2 = 1$. Thus, with a little algebra and the quadratic formula,

$$x_{1,2} = \frac{-mb \pm \sqrt{m^2 + 1 - b^2}}{1 + m^2}.$$

The two intersection points of the line with the *full circle* have y-coordinates given by

$$y_{1,2} = mx_{1,2} + b.$$

Now, as shown in Figure S7.2, there are four possibilities for the location of these two intersection points, with either none, one, or both of them being on the circular arc in the first quadrant. The code determines which of the four possibilities has occurred by examining the products $x_1 x_2$ and $y_1 y_2$. After studying Figure S7.2, you should be able to convince yourself that

For 0 intersections: $x_1 x_2 < 0$ and $y_1 y_2 < 0$;
For 1 intersection: $x_1 x_2 < 0$ and $y_1 y_2 > 0$, or $x_1 x_2 > 0$ and $y_1 y_2 < 0$;
For 2 intersections: $x_1 x_2 > 0$ and $y_1 y_2 > 0$.

When run, **crossings.m** estimated the three probabilities as

Probability of 0 intersections: 0.1350856;
Probability of 1 intersection: 0.8105033;
Probability of 2 intersections: 0.0544111.

This problem did receive a theoretical analysis in the January 1879 *Mathematical Visitor*, with the following results:

Probability of 0 intersections: $\frac{4}{3\pi^2} = 0.13509$;
Probability of 1 intersection: $\frac{8}{\pi^2} = 0.810569$;
Probability of 2 intersections: $1 - \frac{28}{3\pi^2} = 0.05433$.

The Monte Carlo simulation estimates are in excellent agreement with these theoretical values. The clever analyst was the one-time carpenter Henry Heaton (1846–1927), a self-taught mathematician who became a public school administrator in Iowa.

Finally, the code **stack.m** simulates the random stacking of four identical bricks.

```
%stack.m/created by PJNahin for Mathematical Physics
rand('state',100*sum(clock));
topple=0;
L1=0;R1=1;
for n=1:10000000
    L2=(L1-1)+(R1-L1+1)*rand;R2=L2+1;
    L3=(L2-1)+(R2-L2+1)*rand;R3=L3+1;
    L4=(L3-1)+(R3-L3+1)*rand;R4=L4+1;
    CM2=(L2+R2)/2;
    CM3=(L3+R3)/2;
    CM4=(L4+R4)/2;
    CM23=(CM2+CM3)/2;
    CM34=(CM3+CM4)/2;
    CM234=(CM2+CM3+CM4)/3;
        if CM2>R1|CM2<L1
            topple=topple+1;
        elseif CM3>R2|CM3<L2
            topple=topple+1;
```

(continued)

```
                                                          (continued)
        elseif CM23>R1|CM23<L1
            topple=topple+1;
        elseif CM4>R3|CM4<L3
            topple=topple+1;
        elseif CM34>R2|CM34<L2
            topple=topple+1;
        elseif CM234>R1|CM234<L1
            topple=topple+1;
        end
    end
    1-(topple/10000000)
```

After fixing the left and right edges of brick 1 at $L1 = 0$ and $R1 = 1$, then for each of 10 million simulations the code randomly selects the left- and right-edge locations of bricks 2, 3, and 4, subject to the constraint of note 5 in the problem statement. The center of mass for each individual brick is then computed and used to find the center of mass for all the bricks above each brick. The stack remains intact if and only if the center of mass of all the bricks over each brick remains over that brick. When run multiple times, **stack.m** gave estimates for the probability that a random stack remains intact after the fourth brick is added that varied from 0.0905 to 0.0908.

The first theoretical solution was published in the September 1876 issue of *The Analyst*, in a note by the American mathematician and medical doctor Joel Hendricks (1818–1893). Hendricks mentioned that Henry Heaton (mentioned earlier in connection with **crossings.m**) had also found a solution (Hendricks and Heaton were friends), as had the American mathematical academic Enoch Seitz (1846–1883). Later, in the September 1890 issue of *The Mathematical Messenger*, Heaton finally published his solution (but I have had no luck in locating Seitz's[1]). The solutions by Hendricks and Heaton appear to be different, but in the end, they arrive at the same answer:

---

[1] The closest I've come is finding Seitz's solution to what I think is a *much* more difficult stacking problem (both theoretically and to simulate), involving cubic dice.

$\frac{209}{2,304} = 0.0907118\ldots$, in good agreement with **stack.m**. This problem is a good example of the utility of Monte Carlo, as it would be a real challenge to extend either of the theoretical analyses to, say, stacking *five* bricks. To extend **stack.m**, however, would be quite simple, requiring only the addition of a few more center-of-mass and elseif statements.

PROBLEM 8
Monte Carlo *and* Theory

Given $f_N(x) = qf_N(x-1) + pf_N(x+1)$, with $f_N(0) = 0$ and $f_N(N) = 1$, we'll follow the hint and *assume* $f_N(x) = Ce^{sx}$, with C and s constants. Substituting this assumed $f_N(x)$ into the difference equation, we have

$$Ce^{sx} = qCe^{s(x-1)} + pCe^{s(x+1)} = qCe^{sx}e^{-s} + pCe^{sx}e^{s},$$

or dividing out the common Ce^{sx} factor, we arrive at $1 = \frac{q}{e^s} + pe^s$, or $e^{2s} - e^s\frac{1}{p} + \frac{q}{p} = 0$. This is a quadratic in e^s, and so we can immediately write

$$e^s = \frac{\frac{1}{p} \pm \sqrt{\frac{1}{p^2} - 4\frac{q}{p}}}{2} = \frac{1 \pm \sqrt{1 - 4pq}}{2p}.$$

Thus, $s = \ln\left\{\frac{1}{2p}\left(1 \pm \sqrt{1 - 4pq}\right)\right\}$. That is, there are two solutions for s *as long as* $p \neq q$. If $p = q$ ($= \frac{1}{2}$, since $p + q = 1$), then there is just one solution for s, a special case we'll tackle in just a bit. For now, let's

---

In this problem the stacking is allowed to be irregular, that is, not like a brick wall but instead allowing a cube to possibly lie at an angle over a corner of the cube below it. Seitz's answer to the question of the probability of a three-cube stack standing was $\frac{\pi}{(\pi+4)^2}[\frac{\pi}{4} + \frac{1}{96} - \frac{\sqrt{2}}{12} + \frac{1}{12}\log(\sqrt{2} - 1)]$, which you can find derived on p. 157 of the July 1880 issue of *The Mathematical Visitor*.

assume $p \neq q$. Notice that

$$\frac{1 \pm \sqrt{1 - 4pq}}{2p} = \frac{1 \pm \sqrt{1 - 4p(1-p)}}{2p} = \frac{1 \pm \sqrt{1 - 4p + 4p^2}}{2p}$$

$$= \frac{1 \pm \sqrt{(1 - 2p)^2}}{2p} = \frac{1 \pm (1 - 2p)}{2p}.$$

Using the $+$ sign reduces this to $\frac{q}{p}$, while using the $-$ sign gives 1. So, the two values of s are $s_1 = \ln(1) = 0$, and $s_2 = \ln(\frac{q}{p})$. Therefore, $f_N(x) = C_1 e^{s_1 x} + C_2 e^{s_2 x} = C_1 + C_2 e^{x \ln(\frac{q}{p})} = C_1 + C_2 e^{\ln(\frac{q}{p})x} = C_1 + C_2(\frac{q}{p})^x$. Since $f_N(0) = 0$, we have $0 = C_1 + C_2$, or $C_1 = -C_2$. So, dropping the subscripts on the C's, we have $f_N(x) = C - C(\frac{q}{p})^x$. And since $f_N(N) = 1$, we have

$$1 = C - C\left(\frac{q}{p}\right)^N = C\left\{1 - \left(\frac{q}{p}\right)^N\right\},$$

or

$$C = \frac{1}{1 - \left(\frac{q}{p}\right)^N}.$$

Thus,

$$f_N(x) = C - C\left(\frac{q}{p}\right)^x = C\left\{1 - \left(\frac{q}{p}\right)^x\right\},$$

or, at last, for $p \neq q$,

$$\boxed{f_N(x) = \frac{1 - \left(\frac{q}{p}\right)^x}{1 - \left(\frac{q}{p}\right)^N}, \quad x = 0, 1, 2, \ldots, N.}$$

Notice that this reduces, for all x, to the *indeterminate* $\frac{0}{0}$ if $p = q = \frac{1}{2}$ (the so-called symmetrical drunkard's random walk), and we'll need to do one last calculation (in just a bit) for that special case. For the case of $p = 0.6$ (and so $q = 0.4$) and $N = 11$, however, we have

$$f_{11}(7) = \frac{1 - \left(\frac{0.4}{0.6}\right)^7}{1 - \left(\frac{0.4}{0.6}\right)^{11}} = \frac{0.941472}{0.988439} = 0.952,$$

which is the value I gave you in the problem statement. A Monte Carlo code that simulates this walk is quite straightforward (notice the $p = q$ case, which presents a complication in the theoretical analysis, offers none to the simulation) and is given in **walk.m** (in MATLAB, $\sim=$ means "not equal").

```
%walk.m/created by PJNahin for Mathematical Physics
p=0.6;home=0;N=11;
initialx=7;
for loop=1:1000000
    x=initialx;
    while x~=0&&x~=N
        if rand<p
            x=x+1;
        else
            x=x-1;
        end
    end
    if x==N
        home=home+1;
    end
end
home/loop
```

Here's a comparison of $f_{11}(x)$, as computed from the theoretical expression in the first box, and from **walk.m** using the average of several individual 1 million walks for each value of x from 1 to 10.

x	**walk.m**	Theory
1	0.337	0.337
2	0.562	0.562
3	0.712	0.712
4	0.812	0.812
5	0.878	0.878
6	0.923	0.923
7	0.952	0.952
8	0.972	0.972
9	0.985	0.985
10	0.994	0.994

As you can see, the agreement between theory and the Monte Carlo "experiment" is, to three decimal places, *perfect*. Now, what about the loose end of the indeterminate case of $p = q = \frac{1}{2}$? If you're inspired enough, perhaps you've already seen that $f_N(x) = \frac{x}{N}$ works *in that special case*, which you can confirm by direct substitution into the difference equation. That is, such a substitution gives (on the right-hand side of that equation)

$$q\left\{\frac{x-1}{N}\right\} + p\left\{\frac{x+1}{N}\right\} = \frac{x}{N}\{q + p\} + \frac{1}{N}\{p - q\} = \frac{x}{N},$$

which is, indeed, the assumed $f_N(x)$. But what if you're not so inspired? Then you can remove the indeterminacy by remembering a result from freshman calculus: L'Hôpital's rule[1], that is,

$$\text{if } \lim_{r \to a} \frac{g(r)}{h(r)} = \frac{0}{0}, \quad \text{then } \lim_{r \to a} \frac{g(r)}{h(r)} = \frac{\lim \frac{dg}{dr}_{r \to a}}{\lim \frac{dh}{dr}_{r \to a}}.$$

---

[1] Named after the French marquis G.F.A. de L'Hôpital (1661–1704). This result is, however, actually the work of L'Hôpital's teacher, the Swiss mathematician Jean (aka Johann) Bernoulli (1667–1748), who discovered it in 1694. The rule is part of any mathematical physicist's tricks of the trade.

Now, we write $r = \frac{q}{p}$, and so we are interested in the case of

$$\lim_{r \to 1} \frac{1 - r^x}{1 - r^N} = \frac{0}{0}.$$

From L'Hôpital's rule, we have

$$\lim_{r \to 1} \frac{1 - r^x}{1 - r^N} = \lim_{r \to 1} \frac{-x r^{x-1}}{-N r^{N-1}} = \frac{x}{N}.$$

PROBLEM 9
More Monte Carlo

The code **paris.m** simulates the drunken tourist as he wanders about Paris. The connection of the street nodes is described by the *data* statements, where *data(j,:)* details the nodes that are the immediate neighbors of node *j*. A simulation starts with the variable *currentnode* initially set to the value of *startnode* and then, as long as the current node is *not equal* (in MATLAB that's what $\sim=$ means) to either 9 (the Arc) or 17 (the outskirts), the new value for *currentnode* is randomly selected from the immediate neighbors of the present current node. The simulation continues until it is terminated by the tourist reaching either node 17, or node 9 (which causes the variable arc to be incremented).

```
%paris.m/created by PJNahin for Mathematical Physics
data(1,:)=[2,5,6,17,17];
data(2,:)=[1,3,4,17,17];
data(3,:)=[2,4,11,17,17];
data(4,:)=[2,3,5,9,10];
data(5,:)=[1,4,6,8,9];
```

(continued)

(continued)

```
data(6,:)=[1,5,7,17,17];
data(7,:)=[6,8,14,17,17];
data(8,:)=[5,7,9,13,14];
data(9,:)=[4,5,8,10,13];
data(10,:)=[4,9,11,12,13];
data(11,:)=[3,10,12,17,17];
data(12,:)=[10,11,15,17,17];
data(13,:)=[9,8,10,15,16];
data(14,:)=[7,8,15,17,17];
data(15,:)=[13,14,16,17,17];
data(16,:)=[12,13,15,17,17];
startnode=1;arc=0;
for loop=1:10000000
  currentnode=startnode;steps=0;
  while currentnode~=9&currentnode~=17
    R=rand;
    if R<0.2
      currentnode=data(currentnode,1);
    elseif R<0.4
      currentnode=data(currentnode,2);
    elseif R<0.6
      currentnode=data(currentnode,3);
    elseif R<0.8
      currentnode=data(currentnode,4);
    else
      currentnode=data(currentnode,5);
    end
  end
  if currentnode==9
    arc=arc+1;
  end
end
arc/10000000
```

When run with startnode=j, the results were

$$p(1) = p(2) = 0.1429$$
$$p(4) = p(5) = 0.4284$$
$$p(3) = p(6) = 0.1428$$
$$p(7) = p(11) = 0.1429$$
$$p(8) = p(10) = 0.4285$$
$$p(12) = p(14) = 0.1429$$
$$p(15) = p(16) = 0.1428$$
$$p(13) = 0.4283$$

I think it most unlikely that you can look at these numbers and not at least *suspect* that 0.1428(9) is 1/7, and that 0.4283(4)(5) is 3/7. The simulation confirms the equalities I asked you to think about (the reason for them is *symmetry*, in that nodes 1 and 2 are in precisely the same relationship with the other nodes as are 4 and 5, 3 and 6, 7 and 11, 8 and 10, 12 and 14, and 15 and 16). The simulation results further suggest that nodes 1, 2, 3, 6, 7, 11, 12, 14, 15, and 16 are similarly "identical" with each other, as are nodes 4, 5, 8, 10, and 13 with each other. With these suggested relationships, can you now develop a theoretical analysis to *derive* the $p(j)$ probabilities?

PROBLEM 10
Flying with (and against) the Wind

For the warm-up question, the key idea is that the meaning of the word *half* is ambiguous. Does it mean half the *time* of the run, or half the *distance* of the run? Let T be the time of a run. Then, if half refers to *distance*, we have the first 4 miles at 12 miles per hour, and the second 4 miles at 8 miles per hour. Thus, $T = \frac{4}{12} + \frac{4}{8} = \frac{1}{3} + \frac{1}{2} = \frac{5}{6}$ hour, or 50 minutes. However, if half refers to *time*, then we have $\frac{1}{2}T$ at 12 miles per hour, and $\frac{1}{2}T$ at 8 miles per hour. Thus, $\frac{1}{2}T(12) + \frac{1}{2}T(8) = 8$, or $6T + 4T = 8$, or $10T = 8$, or $T = \frac{8}{10} = \frac{4}{5}$ hour, or 48 minutes. For the second half of this problem, we write $\frac{2D}{S}$ as the total roundtrip flight time *when no wind blows at all*. Now, we can write the wind speed from **A** to **B** as $w(t) = kt$ (where k is some constant), and so the plane's ground speed is $v(t) = S + kt$. Now, the differential distance dx traveled by the plane in differential time dt is

$$dx = v(t)dt = (S + kt)dt,$$

and so if T_1 is the total time to fly from **A** to **B**, we have

$$D = \int_0^{T_1} (S + kt)dt = ST_1 + \frac{1}{2}kT_1^2.$$

Since $w(T_1) = W = kT_1$, we have $k = \frac{W}{T_1}$, and so

$$D = ST_1 + \frac{1}{2}\frac{W}{T_1}T_1^2 = T_1\left(S + \frac{1}{2}W\right),$$

or $T_1 = \frac{D}{S + \frac{1}{2}W}$. Since the wind speed on the return trip from **B** to **A** is the constant $\frac{1}{2}W$, the plane's ground speed is $S - \frac{1}{2}W$, and so the return flight time is $T_2 = \frac{D}{S - \frac{1}{2}W}$. Thus, the total roundtrip flight time *when the wind blows* is

$$T_1 + T_2 = \frac{D}{S + \frac{1}{2}W} + \frac{D}{S - \frac{1}{2}W} = \frac{2D}{S}\left[\frac{1}{1 - \frac{1}{4}\left(\frac{W}{S}\right)^2}\right],$$

which is obviously *greater* than $\frac{2D}{S}$. If $W = \frac{1}{2}S$, then, specifically,

$$T_1 + T_2 = \frac{2D}{S}\left(\frac{16}{15}\right) = 1.067\frac{2D}{S}.$$

That is, when the wind blows the roundtrip takes nearly 7% longer than when no wind blows.

Extra-Credit Challenge: We start by writing

$$\left\{\frac{1}{T}\int_0^T v(t)dt\right\}^2 = \frac{1}{T^2}\left\{\int_0^T 1\cdot v(t)dt\right\}^2$$

and then invoke the Cauchy-Schwarz inequality to write

$$\left\{\frac{1}{T}\int_0^T v(t)dt\right\}^2 \leq \frac{1}{T^2}\left\{\int_0^T 1^2 dt \int_0^T v^2(t)dt\right\}$$

$$= \frac{1}{T^2}T\int_0^T v^2(t)dt = \frac{1}{T}\int_0^T v^2(t)dt.$$

Now, since $v(t) = \frac{dx}{dt}$, this becomes

$$\left\{\frac{1}{T}\int_0^T v(t)dt\right\}^2 \leq \frac{1}{T}\int_0^T v(t)\left(\frac{dx}{dt}\right)dt = \frac{1}{T}\int_0^T v(t)dx,$$

or

$$\frac{1}{T}\left\{\int_0^T v(t)dt\right\}^2 \leq \int_0^T v(t)dx.$$

We can replace $v(t)$ with $v(x)$ in the integral on the right if we change the limits on that integral to be the appropriate limits for the integration variable, x. Thus,

$$\frac{1}{T}\left\{\int_0^T v(t)dt\right\}^2 \leq \int_0^D v(x)dx,$$

or

$$\frac{1}{T}\int_0^T v(t)dt \int_0^T v(t)dt \leq \int_0^D v(x)dx.$$

Now, since $\int_0^T v(t)dt = D$, we have

$$\frac{D}{T}\int_0^T v(t)dt \leq \int_0^D v(x)dx,$$

or finally,

$$\frac{1}{T} \int\limits_0^T v(t)dt \leq \frac{1}{D} \int\limits_0^D v(x)dx,$$

as was to be shown.

PROBLEM 11
A Combinatorial Problem with Physics Implications

The answer to the first question is, *immediately*, zero. That's because *if Pauli exclusion applies*, it is impossible to put n balls into n boxes with one box remaining empty, since that would require one box to contain more than one ball. The answer to the second question requires just a bit more thought. First, as agreed in the problem statement, there are n^n distinguishable ways to put n distinguishable balls into n distinguishable boxes. To calculate the number of those ways that leave one box empty, we proceed as follows:

    a. There are n choices for which box will be the empty one;

    b. there are now $n-1$ boxes left to receive n balls;

    c. each of those $n-1$ boxes must get a ball, otherwise there would be more than one empty box;

    d. there are n choices for the ball to go into the first box;

    e. there are $n-1$ choices for the ball to go into the second box;

    f. and so on, until we get to the last (that is, the $n-1$th) box;

    g. the final ball can go into any one of the $n-1$ boxes that already have a ball (remember, Pauli exclusion does *not* apply);

    h. thus, we have $n[(n)(n-1)\cdots(2)](n-1)\frac{1}{2} = \frac{n(n-1)n!}{2}$ ways to put n distinguishable balls into n distinguishable boxes, with exactly one box remaining empty.[1] Thus, the probability of one empty box after

---

[1]The factor of $\frac{1}{2}$ comes from the fact that the one box that has *two* balls in it could have received those two particular balls in either order, and the two orderings are, in the end, *not* distinguishable. That is, $n(n-1)n!$ counts all the *distinguishable* ways *twice*, so, we divide by 2.

putting n distinguishable balls into n distinguishable boxes with Pauli exclusion in effect is

$$\frac{n(n-1)n!}{2n^n} = (n-1)\frac{n!}{2n^{n-1}}.$$

For $n = 4$ this is $3\frac{4!}{2(4^3)} = 0.5625$, while for $n = 7$ it is $6\frac{7!}{2(7^6)} = 0.128517\ldots$, the values given in the problem statement. The code **sevenballs.m** simulates—10 million times—putting seven balls into seven boxes, and keeps track of how many times exactly one box remains empty. At the start of each simulation, all seven boxes are initialized empty: $box(j) = 0$ for $1 \le j \le 7$. Then, if the jth box receives a ball it is set to be *not* empty: $box(j) = 1$. This can happen multiple times for the same box, but after the first time, putting another ball into a nonempty box has no effect on the box's "emptiness"! At the end of a simulation, the sum of all the $box(j)$ is the number of nonempty boxes, so a sum of 6 means exactly one box remains empty (and the variable *empty* is incremented). The Monte Carlo estimate of the probability of that occurring is 0.1285289, in good agreement with theory.

```
%sevenballs.m/created by PJNahin for Mathematical Physics
empty=0;
b1=1/7;b2=2/7;b3=3/7;b4=4/7;b5=5/7;b6=6/7;
for loop=1:10000000
    for j=1:7
        box(j)=0;
    end
    for ball=1:7
        decision=rand;
        if decision<b1
            box(1)=1;
        elseif decision<b2
            box(2)=1;
```

(continued)

(continued)

```
        elseif decision<b3
            box(3)=1;
        elseif decision<b4
            box(4)=1;
        elseif decision<b5
            box(5)=1;
        elseif decision<b6
            box(6)=1;
        else
            box(7)=1;
        end
    end
    summ=box(1)+box(2)+box(3)+box(4)+box(5)+box(6)+box(7);
    if summ==6
        empty=empty+1;
    end
end
empty/loop
```

PROBLEM 12
Mathematical Analysis

Following the hint, for the first challenge we write $\binom{x}{r} = \frac{x(x-1)(x-2)\dots(x-r+1)}{r!}$ for x any real number and r a positive integer. Then,

$$
\begin{aligned}
\binom{-1}{r} &= \frac{(-1)(-2)(-3)\dots(-r)}{r!} \\
&= \frac{[(-1)(1)]\,[(-1)(2)]\,[(-1)(3)]\dots[(-1)(r)]}{r!} \\
&= \frac{(-1)^r r!}{r!} = (-1)^r.
\end{aligned}
$$

Also,

$$
\begin{aligned}
\binom{-2}{r} &= \frac{(-2)(-3)\ldots(-r-1)}{r!} \\
&= \frac{[(-1)(2)]\,[(-1)(3)]\ldots[(-1)(r+1)]}{r!} \\
&= \frac{(-1)^r r!(r+1)}{r!} = (-1)^r(r+1),
\end{aligned}
$$

which establishes the two given identities. Now, in the same way,

$$
\begin{aligned}
\binom{-3}{r} &= \frac{(-3)\ldots(-r-2)}{r!} = \frac{[(-1)(3)]\,[(-1)(4)]\ldots[(-1)(r+2)]}{r!} \\
&= \frac{(-1)^r(3)(4)\ldots(r)(r+1)(r+2)(2)}{r!(2)} = \frac{(-1)^r r!(r+1)(r+2)}{r!(2)},
\end{aligned}
$$

or $\binom{-3}{r} = (-1)^r \frac{1}{2}(r+1)(r+2)$, and so

$$
\boxed{\binom{-3}{r} = (-1)^r \frac{1}{2}(r+1)(r+2).}
$$

So, $\binom{-3}{7} = (-1)^7 \frac{1}{2}(8)(9) = -36$, and $\binom{-3}{8} = (-1)^8 \frac{1}{2}(9)(10) = 45$, the numbers given in the hint.

For the second challenge, and again because $\binom{x}{r} = \frac{x(x-1)(x-2)\ldots(x-r+1)}{r!}$, we have

$$
\frac{\binom{\frac{1}{2}}{n}}{\binom{-\frac{1}{2}}{n}} = \frac{\left(\frac{1}{2}\right)\left(-\frac{1}{2}\right)\left(-\frac{3}{2}\right)\left(-\frac{5}{2}\right)\ldots\left(\frac{3}{2}-n\right)}{\left(-\frac{1}{2}\right)\left(-\frac{3}{2}\right)\left(-\frac{5}{2}\right)\ldots\left(\frac{1}{2}-n\right)},
$$

where there are n factors in the numerator and in the denominator. So,

$$\frac{\binom{\frac{1}{2}}{n}}{\binom{-\frac{1}{2}}{n}} = \frac{\frac{1}{2}(-1)^{n-1}\left[\left(\frac{1}{2}\right)\left(\frac{3}{2}\right)\left(\frac{5}{2}\right)\cdots\left(\frac{2n-3}{2}\right)\right]}{(-1)^n\left[\left(\frac{1}{2}\right)\left(\frac{3}{2}\right)\left(\frac{5}{2}\right)\cdots\left(\frac{2n-3}{2}\right)\left(\frac{2n-1}{2}\right)\right]} = \frac{-\frac{1}{2}}{\frac{2n-1}{2}} = -\frac{1}{2n-1},$$

which is (as claimed) equal to $-\frac{1}{5}$ for $n = 3$.

For the third challenge, following the hint we have

$$\binom{2n}{n} - \binom{2n}{n+1} = \frac{(2n)!}{n!n!} - \frac{(2n)!}{(n+1)!(n-1)!} = \frac{(2n)!}{n!n!} - \frac{(2n)!n}{(n+1)n!n!}$$

$$= \frac{(2n)!}{n!n!}\left[1 - \frac{n}{n+1}\right] = \frac{(2n)!}{n!n!}\left[\frac{n+1-n}{n+1}\right]$$

$$= \frac{1}{n+1}\binom{2n}{n} = C_n.$$

Done.

PROBLEM 13
When an Integral Blows Up

From the hint we have $z^2 = \frac{1-\cos(\theta)}{1+\cos(\theta)}$, which gives $\cos(\theta) = \frac{1-z^2}{1+z^2}$. Differentiating this with respect to z, we get

$$-\sin(\theta)\frac{d\theta}{dz} = -\frac{4z}{\left(1+z^2\right)^2},$$

or

$$\sin(\theta)\frac{d\theta}{dz} = \frac{4z}{\left(1+z^2\right)^2}.$$

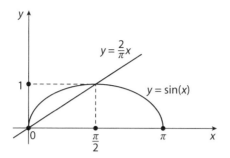

Figure S13.1. Proof without words.

From the result for $\cos(\theta)$ we have (draw a triangle!) $\sin(\theta) = \frac{2z}{1+z^2}$. Using this in the $\sin(\theta)\frac{d\theta}{dz}$ result, we have

$$\frac{2z}{1+z^2}\frac{d\theta}{dz} = \frac{4z}{\left(1+z^2\right)^2},$$

or

$$d\theta = \frac{2}{1+z^2}dz.$$

So,

$$I(\alpha) = \int_0^\alpha \frac{d\theta}{\sqrt{1-\cos(\theta)}} = \int_0^{\tan\left(\frac{\alpha}{2}\right)} \frac{2}{1+z^2}dz\,\frac{1}{\sqrt{1-\frac{1-z^2}{1+z^2}}}$$

$$= 2\int_0^{\tan\left(\frac{\alpha}{2}\right)} \frac{1}{1+z^2}\sqrt{\frac{1+z^2}{2z^2}}dz = \sqrt{2}\int_0^{\tan\left(\frac{\alpha}{2}\right)} \frac{dz}{z\sqrt{1+z^2}}dz.$$

From standard math tables we have

$$\int \frac{dx}{x\sqrt{x^2+a^2}} = \frac{1}{a}\ln\left(\frac{a+\sqrt{x^2+a^2}}{x}\right),$$

and so, with $a = 1$, we have

$$I(\alpha) = \sqrt{2}\left\{\ln\left(\frac{1+\sqrt{x^2+1}}{x}\right)\right\}\Bigg|_0^{\tan\left(\frac{\alpha}{2}\right)},$$

which does, indeed, blow up logarithmically at the lower limit of $x = 0$.

And for the inequality, observe in Figure S13.1 that the line is always *below* $\sin(x)$ over the interval $0 < x < \frac{\pi}{2}$.

PROBLEM 14

Is This Easier Than Falling Off a Log?

The MATLAB code **slide1.m** produces the results shown.

```
%slide1.m/created by PJNahin for Mathematical Physics
R=3;v0=1e-5;g=9.81;
B=1+(v0^2)/(2*g*R);F=sqrt(R/(2*g));
fun=@(x)1./sqrt(B-cos(x));
A=(2/3)+(v0^2)/(3*g*R);
A=acos(A);
tp=F*integral(fun,0,A)
```

*	$v_0 = 10^{-4}$ meters per second	$v_0 = 10^{-5}$ meters per second
$R = 3$ meters	6.3243 seconds	7.5976 seconds
$R = 6$ meters	9.2149 seconds	11.0157 seconds

PROBLEM 15

When the Computer Fails

The following plot shows $P(r, n)$, which displays an interesting "threshold" feature. That is, for $r < 1{,}500$ or so people, P remains essentially zero. Once r gets past 1,600 or so, however, P increases rather rapidly, going above $\frac{1}{2}$ at about $r \approx 2{,}300$ or so people. A detailed numerical

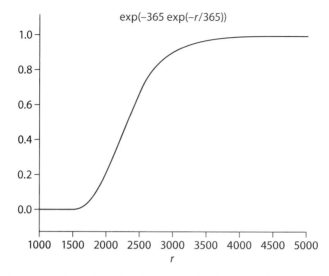

Figure S15.1. Probability that every day in a year is a birthday.

evaluation of P shows that the first value of r, so that $P > \frac{1}{2}$, is $r = 2,288$. ($P = 0.4998$ with $r = 2,287$, and $P = 0.5007$ with $r = 2,288$.) Once r exceeds 4,000 or so, P is very close to 1. It is virtually certain that every day is a birthday in any city with a population in excess of 5,000.

Problem 16
When Intuition Fails

MATLAB code **repeat.m** when run numerous times, each time executing 1 million simulations of drawing (replacing) from a box with 1,000 balls until a repetition, consistently reported an average of 39.3 drawings, in *outstanding* agreement with theory.

```
%repeat.m/created by PJNahin for Mathematical Physics
rand('state',100*sum(clock))
total=0;
```

(continued)

(continued)

```
for loop=1:1000000
    box=zeros(1,1000);duplicate=0;drawing=-1;
    while duplicate==0
        pick=floor(1000*rand)+1;
        drawing=drawing+1;
        if box(pick)==1
            duplicate=1;
        else
            box(pick)=1;
        end
    end
    total=total+drawing;
end
total/loop
```

To find T, the largest number of drawings (after the original drawing) for which the probability of *not* having a repetition is greater than $\frac{1}{2}$, we write

$$\left(\frac{n}{n}\right)\left(\frac{n-1}{n}\right)\left(\frac{n-2}{n}\right)\cdots\left(\frac{n-T}{n}\right) > \frac{1}{2}.$$

To find T, the code **repeat2.m** does the job:

```
%repeatT.m/created by PJNahin for Mathematical Physics
n=input('How many balls?')
j=0;Prod=1;
while Prod>0.5
    Prod=Prod*((n-j)/n);
    j=j+1;
end
T=j-1
```

The results are as follows:

n (number of balls)	T
10	4
100	12
1,000	37
10,000	118
20,000	166
50,000	263

For the final question, plug in the first and fourth rows of the results from the code **balls.m** (refer to Problem 16) into the assumed expression for $E(n)$: $3.66 = k10^x$ and $125 = k10000^x = k10^{4x}$. Then, $\frac{k10^{4x}}{k10^x} = 10^{3x} = \frac{125}{3.66}$, or $x = \frac{1}{3}\log_{10}\left(\frac{125}{3.66}\right)$, or $x = 0.51$. Let's call it 0.5, and so we have a *square root*. That is, $(n) = k\sqrt{n}$. Then, using the fourth row, $k\sqrt{10,000} = 125 = 100k$, or $k = 1.25$. Thus, approximately, $(n) = 1.25\sqrt{n}$. To see how this works on the last row, our formula says that for $n = 50,000$ we should have $E(n) = 1.25\sqrt{50,000} = 279.5$, in excellent agreement with **balls.m**'s result of 280.

PROBLEM 17
Computer Simulation of the Physics of NASTYGLASS

For $b = 11$, we have $2^b = 2,048$, and so

$$\left(2^b\right)^{\left(2^b\right)} = 2,048^{2,048} = (2.048 \times 10^3)^{2,048}$$

$$= 2.048^{2,048} \times 10^{6,144} = (2.048^{128})^{16} \times 10^{6,144}$$

$$= (7.083271604 \times 10^{39})^{16} \times 10^{6,144}$$

$$= 7.083271604^{16} \times 10^{624} \times 10^{6,144} = 4 \times 10^{13} \times 10^{624} \times 10^{6,144}$$

$$= 4 \times 10^{6,781}.$$

Since the area of a circle with radius R inches is πR^2 square inches, there are $100\pi R^2$ grass shoots in the "TUPA lawn." To find R we write $100\pi R^2 = N = 10^{33,113,299}$, and so

$$R = \sqrt{\frac{10^{33,113,297}}{\pi}} = 1.784 \times 10^{16,556,648} \text{ inches.}$$

The number of inches in a light-year is $(186,210)(5,280)(12)$ $(365)(24)(3600) = 3.72 \times 10^{17}$, and so

$$R = \frac{1.784 \times 10^{16,556,648}}{3.72 \times 10^{17}} \text{ light-years} = 4.8 \times 10^{16,556,630} \text{ light-years.}$$

This *vastly* exceeds "hundreds of thousands of millions of billions of trillions of light-years." Finally, there are 80 square inches in an 8-by-10-inch image. So, the pixel area density is $\frac{10,000,000}{80} = 125,000$ pixels per square inch. Thus, the linear density is $\sqrt{125,000} = 353$ pixels per inch. For an image to be published in a mathematical physics journal, the editorial requirement is typically 300 pixels per inch. So, a 10-megapixel, 8-by-10-inch TUPA image is *better* than acceptable.

PROBLEM 18
The Falling Raindrop, Variable-Mass Problem

(a) Following the given steps,

$$\int_0^t \frac{3k}{a+kx} dx = 3k \int_0^t \frac{1}{a+kx} dx.$$

Let $u = a + kx$. Then, $du = kdx$ and so

$$\int_0^t \frac{3k}{a+kx} dx = 3k \int_a^{a+kt} \frac{\frac{du}{k}}{u} = 3\ln(u)\bigg|_a^{a+kt} = 3\ln\left(\frac{a+kt}{a}\right) = \ln\left(\frac{a+kt}{a}\right)^3.$$

So,

$$v(t) = ge^{-\ln\left(\frac{a+kt}{a}\right)^3} \int_0^t e^{\ln\left(\frac{a+ky}{a}\right)^3} dy,$$

or

$$v(t) = g \left(\frac{a+kt}{a} \right)^{-3} \int_0^t \left(\frac{a+ky}{a} \right)^3 dy,$$

or

$$v(t) = g \frac{a^3}{(a+kt)^3} \frac{1}{a^3} \int_0^t (a+ky)^3 dy.$$

Let $u = a + ky$, and so $du = kdy$. Then,

$$\int_0^t (a+ky)^3 dy = \int_a^{a+kt} u^3 \frac{du}{k} = \frac{1}{k} \left(\frac{u^4}{4} \right) \Big|_a^{a+kt}$$

$$= \frac{1}{4k} \left[(a+kt)^4 - a^4 \right].$$

So,

$$v(t) = \frac{g}{4k(a+kt)^3} \left[(a+kt)^4 - a^4 \right] = \frac{g}{4k} \left[a + kt - \frac{a^4}{(a+kt)^3} \right].$$

(b) Since $r(t) = a + kt$, then $kt = r - a$, and so

$$\begin{aligned}
v(t) &= \frac{g}{4k} \left[a + kt - \frac{a^4}{(a+kt)^3} \right] \\
&= \frac{gt}{4kt} \left[\frac{(a+kt)^4 - a^4}{(a+kt)^3} \right] = \frac{gt}{4(r-a)} \left[\frac{r^4 - a^4}{r^3} \right] \\
&= \frac{gt}{4} \left\{ \frac{(r^2-a^2)(r^2+a^2)}{(r-a)r^3} \right\} \\
&= \frac{gt}{4} \left\{ \frac{(r-a)(r+a)(r^2+a^2)}{(r-a)r^3} \right\} = \frac{gt}{4} \left\{ \frac{(r+a)(r^2+a^2)}{r^3} \right\} \\
&= \frac{gt}{4} \left\{ \frac{r^3 + a^2 r + ar^2 + a^3}{r^3} \right\},
\end{aligned}$$

or, as Professor Challis wrote,

$$v(t) = \frac{gt}{4} \left\{ 1 + \frac{a}{r} + \frac{a^2}{r^2} + \frac{a^3}{r^3} \right\}.$$

(c) From the given mass accretion rule, and using results derived in the problem discussion, we have

$$x = \frac{1}{\pi r^2 k} \dot{m} = \frac{1}{\pi r^2 k} \frac{3m}{r} \dot{r} = \frac{3m}{k\pi r^3} \dot{r} = \frac{3m}{k\frac{3m}{4\rho}} \dot{r},$$

or

$$x = \frac{4\rho}{k} \dot{r}.$$

Thus, $\dot{x} = v = \frac{4\rho}{k} \ddot{r}$, and therefore, $\dot{v} = \frac{4\rho}{k} \dddot{r}$. So, as $mg = v\dot{m} + m\dot{v}$, we have

$$mg = \left(\frac{4\rho}{k} \ddot{r}\right) \dot{m} + m \left(\frac{4\rho}{k} \dddot{r}\right),$$

or, as $\dot{m} = \frac{3m}{r} \dot{r}$,

$$mg = \left(\frac{4\rho}{k} \ddot{r}\right) \frac{3m}{r} \dot{r} + m \left(\frac{4\rho}{k} \dddot{r}\right).$$

Thus,

$$g = \frac{4\rho}{k} \left(\frac{3}{r}\right) \ddot{r} \dot{r} + \frac{4\rho}{k} \dddot{r} = \frac{12\rho}{kr} \ddot{r} \dot{r} + \frac{4\rho}{k} \dddot{r},$$

and so

$$\frac{gkr}{\rho} = 12\ddot{r}\dot{r} + 4r\dddot{r}.$$

Now, we *assume* that $r = ct^n$, where c and n are some constants. Then,

$$\frac{gkc}{\rho} t^n = 12 \left\{n(n-1)ct^{n-2}\right\} \left\{nct^{n-1}\right\} + 4ct^n \left\{n(n-1)(n-2)ct^{n-3}\right\},$$

or

$$\frac{gkc}{\rho} t^n = 12c^2 n^2 (n-1)t^{2n-3} + 4c^2 n(n-1)(n-2)t^{2n-3}.$$

For both sides of this equality to have the same t-dependency, we require that $n = 2n - 3$, or $n = 3$. So, assuming this, canceling the t's and dividing through by c, we obtain

$$\frac{gk}{\rho} = 12(9)(2)c + 4c(3)(2)(1)c = 216c + 24c = 240c, \text{ or } = \frac{gk}{240\rho}.$$

Then, using our earlier result of $\dot{v} = \frac{4\rho}{k}\dddot{r}$ (and $n = 3$), we have

$$\dot{v} = \frac{4\rho}{k}(3)(2)(1)C = \frac{24\rho}{k}\left(\frac{gk}{240\rho}\right) = \frac{1}{10}g,$$

and we are done.

(d) The physics of this final part is described by these *assumptions* (as given in the hint): $r = ct^n$, and $\dot{v} = ag$, and so $v = agt$. Thus,

$$m = \frac{4}{3}\pi r^3 \rho = \frac{4}{3}\pi c^3 t^{3n}\rho,$$

and so,

$$\dot{m}\frac{4}{3}\pi c^3 3nt^{3n-1}\rho = 4\pi c^3 3nt^{3n-1}\rho.$$

Since $= v\dot{m} + m\dot{v}$, then

$$\frac{4}{3}\pi c^3 t^{3n}\rho g = agt4\pi c^3 nt^{3n-1}\rho + \frac{4}{3}\pi c^3 t^{3n}\rho ag.$$

Making all the obvious cancellations, we are left with $\frac{1}{3} = an + \frac{1}{3}a$, which says $a = \frac{1}{1+3n}$. To find n, we next invoke the mass accretion rule (which we haven't used yet). That is, $\dot{m} = k\pi r^2 vx$. Since $v = agt = \dot{x}$, then $x = \frac{1}{2}agt^2$ (where I've used $x(0) = 0$). So, substituting into the $\dot{m}$ equation, we have

$$4\pi c^3 nt^{3n-1}\rho = k\pi c^2 t^{2n}agt\frac{1}{2}agt^2.$$

On the left-hand side we have t^{3n-1}, and on the right-hand side we have t^{2n+3}. For consistency we require $3n - 1 = 2n + 3$, or $n = 4$. Putting that into our result for a, we get $a = \frac{1}{13}$, and so $\dot{v} = \frac{1}{13}g$.

PROBLEM 19
Beyond the Quadratic

$$\frac{d^2 U}{dx^2} = c_1 \left[\frac{2q}{x^3} + 1 \right] = c_1 \left[\frac{2\{x^2(x-1)\}}{x^3} + 1 \right] = c_1 \left[\frac{2(x-1)}{x} + 1 \right]$$

$$= c_1 \left[\frac{2x - 2}{x} + 1 \right] = c_1 \frac{3x - 2}{x}.$$

Now, for $x = x_1 > 1$ (see Figure P19.2) we see that $\frac{d^2 U}{dx^2} > 0$, and so $x = x_1$ is a *stable* equilibrium point. For $x = x_2$, where $\frac{2}{3} < x_2 < 1$, we see that $\frac{d^2 U}{dx^2}$ varies from $c_1 \left. \frac{3x-2}{x} \right|_{x=\frac{2}{3}} = 0$ to $c_1 \left. \frac{3x-2}{x} \right|_{x=1} = c_1 > 0$. So, for any x_2 in the interval $\frac{2}{3} < x_2 < 1$ we have $\frac{d^2 U}{dx^2} > 0$, and $x = x_2$ is a *stable* equilibrium point. Finally, for $x = x_3$, where $0 < x_3 < \frac{2}{3}$, we see that $\frac{d^2 U}{dx^2}$ varies from $c_1 \left. \frac{3x-2}{x} \right|_{x=0} = -\infty$ to $c_1 \left. \frac{3x-2}{x} \right|_{x=\frac{2}{3}} = 0$, and we have $\frac{d^2 U}{dx^2} < 0$, and so $x = x_3$ is an *unstable* equilibrium point.

PROBLEM 20
Another Cubic Equation

We have the two spheres *just touching* (as shown in Figure S20.1), with a test mass μ at the "just touching" point. Since the test mass experiences zero net force, we have

$$\frac{GM\mu}{R^2} = \frac{Gm\mu}{r^2},$$

or, as was to be shown, $\frac{r^2}{R^2} = \frac{m}{M}$.

For the second question, we follow the hint that if the two spheres *just touch*, then $d = R + r$. So, from the two given conditions, we have

$$R + r = \frac{d}{1 + \sqrt{\frac{m}{M}}} + \frac{d}{1 + \sqrt{\frac{M}{m}}} = \frac{d + d\sqrt{\frac{M}{m}} + d + \sqrt{\frac{m}{M}}}{1 + \sqrt{\frac{m}{M}} + \sqrt{\frac{M}{m}} + 1}$$

$$= \frac{2d + d\left(\sqrt{\frac{M}{m}} + \sqrt{\frac{m}{M}}\right)}{2 + \sqrt{\frac{m}{M}} + \sqrt{\frac{M}{m}}} = d \frac{2 + \sqrt{\frac{m}{M}} + \sqrt{\frac{M}{m}}}{2 + \sqrt{\frac{m}{M}} + \sqrt{\frac{M}{m}}} = d,$$

the "just touching" condition.

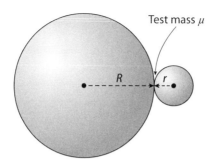

Figure S20.1. Two "just touching" spheres.

Now, before starting the specific calculation of the three planets, consider this preliminary calculation. If we put a test mass μ on the surface of the Earth, then $\mu g = \frac{GM\mu}{R^2}$, where M, g, and R are the mass, surface gravity, and radius of the Earth, respectively. Thus, $g = \frac{GM}{R^2}$.

For the new-planet calculations, let its radius be r and its density equal to Earth's (ρ). Since the Earth's mass is $M = \frac{4}{3}\pi R^3 \rho$, then if the new planet has twice the mass of Earth, we have

$$2M = \frac{4}{3}\pi r^3 \rho = 2\left(\frac{4}{3}\pi R^3 \rho\right),$$

or $r^3 = 2R^3$, or $r = R(2^{1/3})$. Let a be the acceleration of gravity on the surface of the new planet. Thus, a test mass μ on that surface has weight $\mu a = \frac{G2M\mu}{r^2}$, or

$$a = \frac{2GM}{r^2} = 2\frac{GM}{\left\{R\left(2^{1/3}\right)\right\}^2} = 2\frac{GM}{R^2 2^{2/3}},$$

or

$$a = 2^{1/3}\frac{GM}{R^2} = 2^{1/3}g = 1.26g.$$

You could walk on this planet, but you'd notice the required increased effort.

If, however, the new planet has twice the surface area of Earth (which is $4\pi R^2$), then $4\pi r^2 = 2(4\pi R^2)$, or $r = R\sqrt{2}$. The mass of the new planet is therefore

$$m = \frac{4}{3}\pi r^3 \rho = \frac{4}{3}\pi R^3 2\sqrt{2}\rho = 2\sqrt{2}\left(\frac{4}{3}\pi R^3 \rho\right) = 2\sqrt{2}M.$$

So, a test mass μ on the surface of the new planet has weight

$$\mu a = \frac{Gm\mu}{r^2} = \frac{G2\sqrt{2}M\mu}{R^2 2} = \sqrt{2}\mu \frac{GM}{R^2} = \sqrt{2}\mu g,$$

or

$$a = \sqrt{2}g = 1.41g.$$

You could walk on this planet, but probably not for long without becoming exhausted.

Next, suppose the new planet has 6.6 times the mass of Earth, with $r = 1.4R$. A test mass μ on the planet's surface feels the force (has weight)

$$\mu a = \frac{G6.6M\mu}{r^2} = 6.6\frac{GM\mu}{(1.4)^2 R^2},$$

or

$$a = \frac{6.6}{(1.4)^2}\frac{GM}{R^2} = \frac{6.6}{(1.4)^2}g = 3.37g.$$

Thus, an astronaut weighing 150 pounds on Earth would weigh over 500 pounds on the new planet. It would be *impossible* to walk on this planet.

Finally, the energy E_1 required, *against Earth's gravity*, to transport a mass μ from the Earth's surface ($x = R$) to the point of zero gravity ($x = 54R$), is

$$E_1 = \int_R^{54R} F_M(x)dx = \int_R^{54R} \frac{GM\mu}{x^2}dx = GM\mu\left(-\frac{1}{x}\right)\Big|_R^{54R}$$

$$= GM\mu\left(-\frac{1}{54R} + \frac{1}{R}\right) = \frac{53}{54}\frac{GM\mu}{R}.$$

During this journey the Moon's gravitational force is, of course, "helping." Using the fact that the Moon's mass is $m = \frac{1}{81}M$, the energy E_2 provided by the Moon to the mass μ as that mass moves from $x = R$ to $x = 54R$ is

$$E_2 = -\int_{59R}^{6R} F_m(x)dx = -\int_{59R}^{6R} \frac{G\frac{M}{81}\mu}{x^2}dx$$

$$= -G\frac{M}{81}\mu\left(-\frac{1}{x}\right)\Big|_{59R}^{6R} = \frac{GM\mu}{R}\frac{\frac{1}{6}-\frac{1}{59}}{81}.$$

Thus,

$$\frac{E_2}{E_1} = \frac{\frac{\frac{1}{6}-\frac{1}{59}}{81}}{\frac{53}{54}} = \frac{54}{53}\left(\frac{1}{81}\right)\left(\frac{1}{6}-\frac{1}{59}\right) = 0.00188.$$

That is, E_2 is less than 0.2% of E_1. The "aid" of the Moon is insignificant.

Problem 21
Beyond the Cubic

Starting with $x^4 + ax^3 + bx^2 + ax + 1 = 0$, dividing through by x^2 gives $x^2 + ax + b + a\frac{1}{x} + \frac{1}{x^2} = 0$. Now, defining $y = x + \frac{1}{x}$, we have $y^2 = x^2 + 2 + \frac{1}{x^2}$, and so $y^2 - 2 = x^2 + \frac{1}{x^2}$, or $y^2 - 2 + a\left(x + \frac{1}{x}\right) + b = 0$, and so $y^2 + ay + b - 2 = 0$. From the quadratic formula we then have

$$y = \frac{-a \pm \sqrt{a^2 - 4(b-2)}}{2} = p = x + \frac{1}{x}.$$

So, $x^2 + 1 = xp$, or $x^2 - xp + 1 = 0$, and thus, again by the quadratic formula,

$$x = \frac{p \pm \sqrt{p^2 - 4}}{2}, \quad \text{where } p = \frac{-a \pm \sqrt{a^2 - 4(b-2)}}{2},$$

which represents four solutions when the $\pm$ signs are used in all possible combinations.

PROBLEM 22
Escaping an Atomic Explosion

The codes **EG.m** and **shock.m** each have a line containing the factor 1200 (the shock wave speed). Changing that factor to 2400, and then to 3600, produces the following two plots (for the optimal turning angle, which you'll notice remains unchanged) which put the *Enola Gay* at distances of about 49,500 feet and 46,500 feet (slant range), respectively, from the detonation.

PROBLEM 23
"Impossible" Math Made Easy

(a) We start with $88 = 8 \times 11 = 2^3 \times 11 \equiv -1 (\mathrm{mod}\, 89)$. So, squaring, $2^6 \times 11^2 \equiv 1 (\mathrm{mod}\, 89)$. Now, since $11^2 = 121 \equiv 32 (\mathrm{mod}\, 89) = 2^5 (\mathrm{mod}\, 89)$, that is, $11^2 \equiv 2^5 (\mathrm{mod}\, 89)$, we have $2^6 \times 2^5 \equiv 1 (\mathrm{mod}\, 89)$, or $2^{11} \equiv 1 (\mathrm{mod}\, 89)$. Raising this to the fourth power, we have $2^{44} \equiv 1 (\mathrm{mod}\, 89)$, and so $2^{44} - 1$ *is* divisible by 89, as was to be shown.

(b) Next, we write $96 = 32 \times 3 = 2^5 \times 3 \equiv -1 (\mathrm{mod}\, 97)$. So, squaring, $2^{10} \times 9 \equiv 1 (\mathrm{mod}\, 97)$, and then squaring again, $2^{20} \times 81 \equiv 1 (\mathrm{mod}\, 97)$. Since $81 \equiv -16 (\mathrm{mod}\, 97) = -2^4 (\mathrm{mod}\, 97)$, we can write $2^{20} \times (-2^4) \equiv 1 (\mathrm{mod}\, 97)$, and so $-2^{24} \equiv 1 (\mathrm{mod}\, 97)$. That is, $2^{24} \equiv -1 (\mathrm{mod}\, 97)$. Squaring, $2^{48} \equiv 1 (\mathrm{mod}\, 97)$, and so $2^{48} - 1$ *is* divisible by 97, as was to be shown.

(c) Finally, first notice that

$$
\begin{aligned}
1^5 &= 1 \equiv 1 (\mathrm{mod}\, 4) \\
2^5 &= 32 \equiv 0 (\mathrm{mod}\, 4) \\
3^5 &= 243 \equiv 3 (\mathrm{mod}\, 4) \\
4^5 &= 1{,}024 \equiv 0 (\mathrm{mod}\, 4)
\end{aligned}
$$

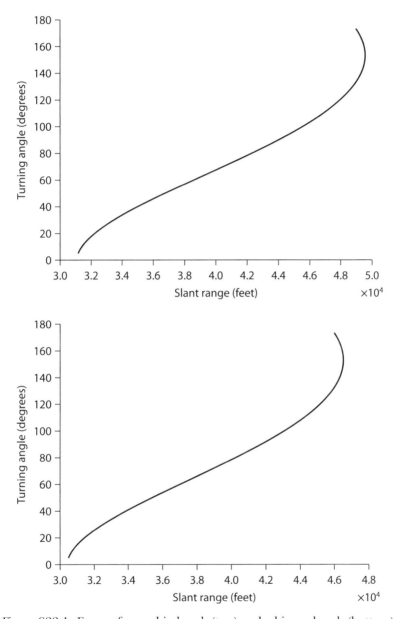

Figure S22.1. Escape from a big bomb (top) and a bigger bomb (bottom).

Then notice that

$$1 \equiv 5 \equiv 9 \equiv \dots \pmod 4$$
$$2 \equiv 6 \equiv 10 \equiv \dots \pmod 4$$
$$3 \equiv 7 \equiv 11 \equiv \dots \pmod 4$$
$$4 \equiv 8 \equiv 12 \equiv \dots \pmod 4.$$

That is, for each group of 4 (1 to 4, 5 to 8, 9 to 12, 13 to 16, and so on), for a total of 25 groups from 1 to 100, we get a remainder of 1 from the first number in each group, and a remainder of 3 from the third number. So, the total remainder from each group is 4, giving a total remainder from all 25 groups of 100, which is evenly divisible by 4. That is, the remainder when we divide the sum of the fifth powers of the integers from 1 to 100 by 4 is zero.

PROBLEM 24
Wizard Math

(a) As stated in the text, the average of any $f(x)$ over any interval is the area bounded by $f(x)$, divided by the interval. Since the *unit* impulse bounds, by definition, *unit* area, the average of $\delta(x)$ is $\frac{1}{2\pi}$ over the interval $-\pi < x < \pi$. We get the average of $\delta(x)$ from the area bounded, and not by looking at specific values of $\delta(x)$.

(b) $f(x) = \begin{array}{l} 0, -\pi < x < 0 \\ \pi, 0 < x < \pi \end{array}$. Thus, $a_0 = \frac{1}{\pi} \int_{-\pi}^{\pi} f(x)dx = \frac{1}{\pi} \int_0^{\pi} \pi dx = \pi$.

Also,

$$a_n = \frac{1}{\pi} \int_{-\pi}^{\pi} f(x)\cos(nx)dx$$

$$= \frac{1}{\pi} \int_0^{\pi} \pi \cos(nx)dx = \left[\frac{\sin(nx)}{n} \right]\Big|_0^{\pi} = 0 \text{ for all } n \geq 1.$$

And

$$b_n = \frac{1}{\pi} \int_{-\pi}^{\pi} f(x) \sin(nx) dx = \frac{1}{\pi} \int_{0}^{\pi} \pi \sin(nx) dx$$

$$= \left[-\frac{\cos(nx)}{n} \right]\Big|_{0}^{\pi} = \frac{1}{n}[1 - \cos(n\pi)] = \begin{cases} \frac{2}{n}, & n \text{ odd} \\ 0, & n \text{ even}. \end{cases}$$

That is, $f(x) = \frac{\pi}{2} + 2 \sum_{n=1,3,5,\dots}^{\infty} \frac{\cos(nx)}{n}$.

Now, $\frac{1}{2\pi} \int_{-\pi}^{\pi} f^2(t) dt = \frac{1}{2\pi} \int_{0}^{\pi} \pi^2 dt = \frac{\pi^3}{2\pi} = \frac{1}{2}\pi^2$. Since $\left(\frac{1}{2}a_0\right)^2 + \frac{1}{2}\sum_{n=1}^{\infty}(a_n^2 + b_n^2) = \frac{\pi^2}{4} + \frac{1}{2}\sum_{n \text{ odd}}^{\infty} \frac{4}{n^2}$, then Parseval's theorem says $\frac{1}{2}\pi^2 = \frac{1}{4}\pi^2 + 2\sum_{n \text{ odd}}^{\infty} \frac{4}{n^2}$, or

$$\boxed{\sum_{n \text{ odd}}^{\infty} \frac{1}{n^2} = \frac{1}{8}\pi^2.}$$

We have

$$\boxed{\zeta(2) = \sum_{n=1}^{\infty} \frac{1}{n^2} = \sum_{n \text{ even}}^{\infty} \frac{1}{n^2} + \sum_{n \text{ odd}}^{\infty} \frac{1}{n^2}.}$$

Since

$$\sum_{n \text{ even}}^{\infty} \frac{1}{n^2} = \frac{1}{2^2} + \frac{1}{4^2} + \frac{1}{6^2} + \dots = \frac{1}{(2 \times 1)^2} + \frac{1}{(2 \times 2)^2} + \frac{1}{(2 \times 3)^2} + \dots$$

$$= \frac{1}{2^2}\left[\frac{1}{1^2} + \frac{1}{2^2} + \frac{1}{3^2} + \dots\right] = \frac{1}{4}\zeta(2),$$

then from the second boxed equation we have

$$\zeta(2) = \frac{1}{4}\zeta(2) + \sum_{n\,\text{odd}}^{\infty} \frac{1}{n^2}$$

or, using the result in the first box,

$$\frac{3}{4}\zeta(2) = \frac{1}{8}\pi^2,$$

and so,

$$\zeta(2) = \frac{4}{3}\left(\frac{1}{8}\pi^2\right) = \frac{\pi^2}{6},$$

just as Euler found.

(c)

$$a_0 = \frac{1}{\pi}\int_{-\pi}^{\pi} x^2 dx = \frac{2}{3}\pi^2,$$

$$
a_n = \frac{1}{\pi}\int_{-\pi}^{\pi} x^2\cos(nx)dx = \frac{1}{\pi}\left[\frac{2x}{n^2}\cos(nx) + \left(\frac{x^2}{n} - \frac{2}{n^3}\right)\sin(nx)\right]\Big|_{-\pi}^{\pi}
$$

$$
= \frac{1}{\pi}\left[\frac{2\pi}{n^2}\cos(n\pi) + \frac{2\pi}{n^2}\cos(n\pi)\right] = \frac{4}{n^2}\cos(n\pi),
$$

$b_n = 0$ because $f(x)$ is even. Thus,

$$x^2 = \frac{1}{3}\pi^2 + 4\sum_{n=1}^{\infty} \frac{\cos(n\pi)}{n^2}\cos(nx) = \frac{1}{3}\pi^2 + 4\sum_{n=1}^{\infty} \frac{(-1)^n}{n^2}\cos(nx).$$

Then, integrating (with respect to x) indefinitely,

$$\frac{1}{3}x^3 = \frac{1}{3}\pi^2 x + 4\sum_{n=1}^{\infty} \frac{(-1)^n}{n^3}\sin(nx) + C,$$

where C is the constant of integration. Setting $x = 0$, and since $\sin(nx) = 0$ for all n at $x = 0$, we see that $C = 0$. So,

$$\frac{1}{3}x^3 = \frac{1}{3}\pi^2 x + 4\sum_{n=1}^{\infty}\frac{(-1)^n}{n^3}\sin(nx).$$

Next, setting $x = \frac{\pi}{2}$, we get

$$\frac{\pi^3}{24} = \frac{\pi^3}{6} + 4\sum_{n=1}^{\infty}\frac{(-1)^n}{n^3}\sin\left(n\frac{\pi}{2}\right) = \frac{\pi^3}{6} + 4\left[-\frac{1}{1^3} + \frac{1}{3^3} - \frac{1}{5^3} + \frac{1}{7^3} - \cdots\right],$$

or

$$4\left[\frac{1}{1^3} - \frac{1}{3^3} + \frac{1}{5^3} - \frac{1}{7^3} + \cdots\right] = \frac{\pi^3}{6} - \frac{\pi^3}{24} = \frac{3\pi^3}{24} = \frac{\pi^3}{8}$$

and so, at last,

$$\frac{1}{1^3} - \frac{1}{3^3} + \frac{1}{5^3} - \frac{1}{7^3} + \cdots = \frac{\pi^3}{32}.$$

(d) We have, from the given Fourier series, that $a_0 = 0$, $a_n = -\frac{1}{n}$ for $n \geq 1$, $b_n = 0$. Thus,

$$\frac{1}{2\pi}\int_{-\pi}^{\pi}\ln^2\left\{\left|2\sin\left(\frac{x}{2}\right)\right|\right\}dx = \frac{1}{\pi}\int_{0}^{\pi}\ln^2\left\{2\sin\left(\frac{x}{2}\right)\right\}dx$$

$$= \frac{1}{2}\sum_{n=1}^{\infty}a_n^2 = \frac{1}{2}\sum_{n=1}^{\infty}\frac{1}{n^2} = \frac{1}{2}\left(\frac{\pi^2}{6}\right) = \frac{\pi^2}{12}.$$

So,

$$\int_{0}^{\pi}\ln^2\left\{2\sin\left(\frac{x}{2}\right)\right\}dx = \frac{\pi^3}{12}.$$

PROBLEM 25
The Euclidean Algorithm

Let $d = \gcd(a, b, c)$. That is, d is the *largest* (greatest) number that divides all three of a, b, and c. Now, we define $f = \gcd(a, b)$ and $g = \gcd(f, c)$. The last definition means g divides f, while the first definition means f divides a and b. (All the factors of g are factors of f, all the factors of f are factors of a, and all the factors of f are factors of b.) But that means g divides a and b. It is also true, from $g = \gcd(f, c)$, that g divides c. That is, g divides a, b, and c. Now, since d is the *greatest* common divisor of a, b, and c, it then follows that

$$\boxed{g \leq d.}$$

Next, as observed in the second sentence, d divides a, b, and c. Since $f = \gcd(a, b)$, then d certainly divides f. And, as already stated, d divides c. Since the definition $g = \gcd(f, c)$ says g is the *greatest* common divisor of a, b, and c, it then follows that

$$\boxed{d \leq g.}$$

The two boxed expressions can *both* be true *only* if $d = g$. That is,

$$d = \gcd(\gcd(a, b), c).$$

PROBLEM 26
One Last Quadratic

Since, by definition, there is no current in G (we have intentionally adjusted b and c to achieve this condition), then the current in the good cable is (by Ohm's law) $\frac{E}{c+a}$, and the current in the bad cable is (by Ohm's law) $\frac{E}{b+x}$. And, again by Ohm's law, this means the voltage P is $\frac{E}{c+a}a$, while the voltage Q is $\frac{E}{b+x}x$. Now, here's the crucial observation: for the current in G to be zero it must be the case that P = Q. That is, $\frac{E}{c+a}a = \frac{E}{b+x}x$, and now you see why the precise value of E is unimportant: whatever it is, it cancels away. Solving for x, we have this simple result for what is called a *balanced bridge*: $x = \frac{ab}{c}$. Since we know b and c from our adjustments of the precision resistor boxes, as well as the total resistance a of the good cable, we therefore have a good estimate for x, which tells us just where to dig up the bad cable.

Appendix 1

MATLAB, Primes, Irrationals, and Continued Fractions

A prime number is a positive integer that can't be factored into the product of smaller integers. The first few primes are 2, 3, 5, 7, 11, 13, 17, 19, 23, ..., where the ellipsis dots stand for the words "and so on." By convention, 1 is not a prime (it makes stating some general theorems easier if 1 is excluded). All the other integers are said to be *composite*, as they can be factored (*uniquely* factored, in fact, but I won't prove that here) into the product of primes. For example, $6 = 3 \times 2$ and $8 = 2 \times 2 \times 2$. Notice that 2 is the only even prime; all other primes are odd.

The primes form an infinite set, that is, there is no *largest* prime. No matter how big is the largest *known* prime, there will always be another one larger waiting to be discovered. The famous proof of this claim dates to centuries before Christ, and is usually attributed to Euclid (ca. 350 BC).[1] It is the gold standard among mathematicians for beauty, brevity, and an immediate ability to convince all who see it. The proof is one by contradiction, a technique beloved by mathematicians and one used often in this book. That is, let's assume the opposite of what we wish to prove, and then from that assumption we'll derive

---

[1] Euclid is, of course, known to generations of high school geometry students as the author of *Elements*, a collection of some of the oldest Greek writings on mathematics. *Elements* is probably one of the most widely read texts ever published, second only to the Bible; well over 1,000 editions of it have been published since its original 1482 appearance as a printed book.

a contradiction. That shows our starting assumption—there *is* a largest prime—must be wrong, and so in fact there is no largest prime.

With that *assumption* of a largest prime, the primes are a *finite* set, so let's label them p_1, p_2, p_3, ..., p_n, where $p_1 = 2$, $p_2 = 3$, and so on, up to p_n, which is our assumed last and largest prime. Now, form the number M, where

$$M = p_1 p_2 p_3 \ldots p_n + 1.$$

Clearly, M is not divisible by any of the prime numbers in the assumed finite set of primes, as we always get a remainder of 1 when we divide M by any of those primes. There are therefore only two possible conclusions: (1) M, which is obviously larger than p_n, is itself a prime, or (2) there is a prime number *not* in our assumed finite set (and so larger than p_n) that *does* divide M. In either case, we arrive at the conclusion that there is a prime number that is larger than p_n, in contradiction to our starting assumption of a finite number of primes. So, that assumption must be wrong, and there are an *infinite* number of primes.

While there are an infinite number of primes, they do seem to "thin out" as we proceed to ever-larger numbers. There are, for example, 25 primes less than 100, 168 primes less than 1,000, 1,229 primes less than 10,000, and 9,592 primes less than 100,000. So, the *density* of the primes decreases from 0.25 to 0.168 to 0.1229 to 0.09592 to ... as we go to increasingly higher numbers. Indeed, it is an easy arithmetic proof to show, for *any* value of n, no matter how large, that there are, somewhere, two *consecutive* primes that are separated by a gap of length n (that is, by a sequence of n composite numbers). To see that this is so for any positive n, consider the integers $(n + 1)! + 2$, $(n + 1)! + 3$, ..., $(n + 1)! + (n + 1)$. The first is divisible by 2, the second is divisible by 3, and so on, to the last integer in the string that is divisible by $n + 1$. So, we have n consecutive *composite* integers.

Going to the other extreme, how close can consecutive primes be? The answer is immediately obvious: the minimum gap is 1, as with 2 and 3. Once that pair of primes is passed, however, the minimum gap is 2, as with 3 and 5. That's because given any prime greater than 2, that prime must be odd, and so the very next integer would be even and thus not a prime. Pairs of primes separated by a gap of 2 are called *twin primes* and, unlike Euclid's proof of the infinity of the primes, there is no known proof showing that the twin primes are infinite in number. Mathematicians are fairly sure they *do* form an infinite set but haven't been able to prove it. This question is one of the great unsolved problems in modern mathematics, one particularly notable because it is one that even nonmathematicians can easily understand, and yet even the greatest mathematicians have failed to resolve.[2]

Determining if a given integer is prime is a problem with both pure intellectual fascination and practical application (in cryptography), and, conceptually, it is easy to solve. To see if any particular n is prime, just try to divide it evenly by every integer from 2 on up. That can get pretty tedious if n is large, of course, but it *will* work. We don't have to actually try all the potential divisors, however, and here's why. Suppose $n > 1$ is composite and so has factors p and q. That is,

$$n = pq,$$

where $1 < p < n$, and $1 < q < n$. Now, either $p = q$ (if n is a perfect square), or one of p and q is the larger factor; let's write $p \leq q$ to cover both possibilities. Then, multiplying through this inequality by p we get

$$p^2 \leq pq = n,$$

and so

$$p \leq \sqrt{n}.$$

So, if n is composite, it must have a factor that is no larger than $\sqrt{n}$.

---

[2] In 2013 the University of New Hampshire mathematician Yitang Zhang took a major step forward. He showed that there is an infinity of prime pairs (not necessarily *consecutive* primes) separated by a gap of less than 70 million. Since Zhang's proof, that large (but finite) gap has been greatly reduced; *if* the gap could be reduced to 2, *then* we'd have a proof of the infinity of the twin primes.

There is a systematic procedure called the *sieve of Eratosthenes*[3] for checking the primality of all the integers up to some predetermined maximum integer. For large integers there *are* more powerful methods, but they are also far more technically involved. The sieve is, I think, conceptually beautiful, and it relies totally only on *very* simple arithmetic. In particular, there is *no division*. Here's how it works. Write down a list of all the integers from 2 to N. The first number, 2, is of course a prime. Cross out 2, and all its multiples, on the list. The first uncrossed number remaining on the list (3) is a prime. Next, cross out that number, and all its multiples, on the list (some of which will have already been removed in the previous step). Keep doing this (marking as a prime the first uncrossed number remaining on the list) until the entire list has been so processed. The uncrossed numbers remaining at the end are the primes from 2 to N.

The sieve algorithm is easy to program on a computer, and what follows is a MATLAB code that does the job. This code uses only commands I believe are available in just about any modern scientific programming language. The code is fast, too. When I ran it on my quite ordinary laptop (four years old at the time and not particularly fast even when new) it found all the 5,761,455 primes in the first 100 million integers in less than 5 minutes (just over 279 *seconds*). And that's *without* taking advantage of the speed enhancement that could be achieved by not going beyond $\sqrt{n}$ as a possible factor of n. Here's how **sieve.m** works in finding, for example, all the primes up to 500.

```
% sieve.m/created by PJNahin for Mathematical Physics
01    L=500; k=0;
02    for i=2 : L
03        N(i)=i;
04    end
05    for i=2 : L
06        if N (i) > 0
```
                                                    (continued)

---

[3] Named after a Greek mathematician who lived in the third century BC, a century after Euclid.

(continued)

```
07          k = k + 1; P(k) = i;
08          for j = i : i : L
09              N(j) = -1;
10          end
11      else
12          end
13  end
14  P
```

P =

2	3	5	7	11	13	17	19	23	29	31	37
41	43	47	53	59	61	67	71	73	79	83	89
97	101	103	107	109	113	127	131	137	139	149	151
157	163	167	173	179	181	191	193	197	199	211	223
227	229	233	239	241	251	257	263	269	271	277	281
283	293	307	311	313	317	331	337	347	349	353	359
367	373	379	383	389	397	401	409	419	421	431	433
439	443	449	457	461	463	467	479	487	491	499	

In the following explanation, the line numbers at the far left are *not* part of MATLAB but serve only as convenient references for discussion. The code **sieve.m** creates two lists of numbers, N and P. The values of the numbers are $N(i) = i$ (that is, the ith entry in the N list is the number i), and $P(k) = k$th prime. For example, $N(17) = 17$, and $P(4) = 7$. In the code listing, line 01 sets the two variables L and k to the values 500 and 0, respectively. L is the largest integer in the interval 2 to L that the code will search for primes, and the value of k is the number of primes found at any point during that search (which is zero, of course, at the start of the search). Lines 02, 03, and 04 simply enter all the integers from 2 to L into the N list. You'll notice that as the search begins, $N(i) > 0$ for all i. To understand how the rest of the code works, the final thing you need to know is that the code "crosses out" a number on the N list by setting it to -1, which means that from that point on $N(i) < 0$ if the ith number has been crossed out.

Lines 05 through 13 perform the search. Starting with the loop variable $i = 2$, line 06 will obviously be satisfied because $N(2) = 2 > 0$, and so line 07 increments k to 1 and sets $P(1) = 2$. That is, the code has found its first prime. Lines 08, 09, and 10 then cross out $N(2)$ and all the multiples of 2 up through L (that is, 500). The loop variable i is then incremented to 3, and the same sequence of events occurs; that is, the code sets $P(2) = 3$, the second prime. The loop variable i is then incremented to 4, and now the test in line 06 fails because $N(4)$ was set to -1 when all the multiples of 2 were crossed out. The code then immediately increments the loop variable i to 5, the test in line 06 is satisfied, $P(3)$ is set to 5 (the third prime), and so on until the entire N list has been examined. At the end, line 14 prints the P list. (Please note that when I ran the code for finding all the primes in the first 100 million integers, I first deleted line 14, which saved a *lot* of paper from spewing out of my printer.)

Computer Challenge Problem: The Diophantine equation $y^2 - 17 = x^3$ is known to have several *integer* solutions: the first five positive ones are $x = 2, 4, 8, 43,$ and 52. All these values for x result in integer values for y (try them). Write a computer program to find the next larger integer x that gives an integer value to y. There *is* such a value, but you'll never find it by hand. (*Hint*: It's less than 10,000 but still pretty big.) If you get stuck, or don't have a computer handy, you'll find the solution in a shaded box at the end of this appendix. Okay, back to our discussion of the primes, which are a vast source of fascination and mystery.

To give you just a hint of what I mean by that—as well as to show you mathematical questions that computers *can't* answer and that yield only to *pure thought*—it has been known since medieval times that the so-called harmonic series (the sum of the reciprocals of all the positive integers) blows up. That is,

$$S = \frac{1}{1} + \frac{1}{2} + \frac{1}{3} + \frac{1}{4} + \frac{1}{5} + \frac{1}{6} + \frac{1}{7} + \frac{1}{8} + \frac{1}{9} + \frac{1}{10} + \frac{1}{11} + \ldots = \infty.$$

This often surprises even college graduates (much less high school students) who haven't taken much math, since each new term in the sum is smaller than the previous term. That is, in fact, a *necessary* condition for the sum of an infinite series of all positive terms if it is to have a finite value, but as the harmonic series illustrates, it is not a *sufficient* condition. Here's a simple arithmetic demonstration of the blow-up, due to the French Bishop of Lisieux, Nicholas Oresme (1323–1382):

$$S = 1 + 1/2 + (1/3 + 1/4) + \left(1/5 + 1/6 + 1/7 + 1/8\right) + \ldots$$
$$> 1 + \frac{1}{2} + \left(\frac{1}{4} + \frac{1}{4}\right) + \left(\frac{1}{8} + \frac{1}{8} + \frac{1}{8} + \frac{1}{8}\right) + \ldots$$
$$> 1 + 1/2 + 1/2 + 1/2 + \ldots = \infty,$$

where we endlessly replace each new subsequence of terms (after the first two terms) with length $2^k, k \geq 1$ in the original series, with a *smaller* subsequence that sums to $\frac{1}{2}$. Thus, a *lower* bound on S is infinity, and so $S = \infty$. That's it.[4]

The divergence of the harmonic series is quite slow, and that has some practical consequences that are perhaps not immediately obvious. As a dramatic example consider the practice of recording the total annual snowfall at a ski resort. Every year in which the total is the largest so far observed is labeled a *record year*. A natural question to then ask is, if we measure the total annual snowfall in each of N consecutive years, how many record years should we "expect" to see? (I'll tell you what "expect" means in just a moment.)

To cast this problem in the form of an amusing tale, imagine that each year the snowfall is generated by a minor weather god who has a stack of cards piled on his desk (of course, the weather god might also be female). Each card has a single number written on it, indicating a total snowfall. Each year he takes the card on top of the pile, reads the number on it, and that is that year's snowfall. He then tosses the card away. Where did the pile of N cards come from? They were created, just before the start of the stretch of N years, by the weather god's assistant, who simply wrote a number at random on each of

---

[4]For a physics application of the divergence of the harmonic series, see my book *In Praise of Simple Physics*, Princeton University Press, 2016, pp. 95–97.

the N cards and then, stacking the cards neatly but in no particular order, placed the pile on his boss's desk. From this not particularly detailed description, it is astonishingly true that we can now calculate the "expected" number of record years. Here's how.

Let's call the numbers written on the cards, in ascending order, n_1, n_2, n_3, ..., n_N, which means

$$n_1 < n_2 < n_3 < \ldots < n_N.$$

Since the cards were placed in the pile at random, we don't know the order in which the weather god will read the cards. We assume, in fact, that all possible orderings are equally likely. This corresponds to the assumption that the total snowfall each year is a totally random event.

If $N = 1$, how many record years will we see? The answer is obvious: the *first* year is *always* a record year. What if $N = 2$? Now, things are just a bit more complicated (but not by very much). The pile of cards on the weather god's desk consists of just two cards, with n_1 and n_2 written on them, so there are two possible sequences, each with probability $\frac{1}{2}$: $n_1 n_2$ and $n_2 n_1$. The first sequence represents 2 record years (since $n_1 < n_2$), while the second sequence represents 1 record year. Mathematicians call a quantity that takes on integer-only values with various probabilities a *discrete random variable* and define its *expected value* as the sum of the products of the different values with their respective probabilities. So, over 2 years, we "expect" to see

$$2\left(\frac{1}{2}\right) + 1\left(\frac{1}{2}\right) = 1 + \frac{1}{2}$$

record years.

What if $N = 3$? Now there are six possible sequences (each with probability $\frac{1}{6}$) in which the weather god might draw cards from a particular pile of three cards:

$$n_1 n_2 n_3, n_1 n_3 n_2, n_2 n_1 n_3, n_2 n_3 n_1, n_3 n_1 n_2, n_3 n_2 n_1.$$

The first sequence represents 3 record years, the next three sequences represent 2 record years, and the last two sequences represent 1 record year. So, the expected number of record years is

$$3\left(\frac{1}{6}\right) + 2\left(\frac{3}{6}\right) + 1\left(\frac{2}{6}\right) = \frac{1}{2} + 1 + \frac{1}{3} = 1 + \frac{1}{2} + \frac{1}{3}.$$

You can now probably see a developing pattern here, but let's just run through, briefly, the next case, $N = 4$.

For $N = 4$ there are 24 possible sequences (there are $N!$ possible sequences, in general), and if you list them all, you'll find there is one giving 4 record years ($n_1 n_2 n_3 n_4$), six giving 3 record years (such as $n_2 n_3 n_4 n_1$), eleven giving 2 record years (such as $n_2 n_4 n_3 n_1$), and six giving 1 record year (such as $n_4 n_3 n_1 n_2$). So, for $N = 4$ the expected number of record years is

$$1\left(\frac{6}{24}\right) + 2\left(\frac{11}{24}\right) + 3\left(\frac{6}{24}\right) + 4\left(\frac{1}{24}\right)$$

$$= \frac{6 + 22 + 18 + 4}{24} = \frac{50}{24} = \frac{25}{12} = \frac{12 + 6 + 4 + 3}{12}$$

$$= 1 + \frac{1}{2} + \frac{1}{3} + \frac{1}{4}.$$

In general, if $N = n$, the expected number of record years is

$$1 + \frac{1}{2} + \frac{1}{3} + \frac{1}{4} + \ldots + \frac{1}{n}$$

and this so-called partial sum of the harmonic series increases *very slowly* with increasing n. For example, to first exceed 6 we need $n = 227$, and to further increase the sum to first exceed 7 we have to increase n to 616. Thus, the occurrence of a record year in snowfall at the ski resort is *not* a common event, and we can say that because the harmonic series diverges very slowly. So here we have a surprising connection between a subtle, abstract math phenomenon and the real weather at a ski resort.

Here's another curious fact about the harmonic series: as you add in more and more terms, the partial sums continually grow (to infinity, as we just showed) but *never*, at any point, will a partial sum happen to equal an integer. That is, if we write the partial sum for the first n

terms (where $n \geq 1$),

$$S_n = 1 + \frac{1}{2} + \frac{1}{3} + \frac{1}{4} + \frac{1}{5} + \frac{1}{6} + \frac{1}{7} + \frac{1}{8} + \ldots + \frac{1}{n},$$

there is no n such that S_n happens to be an integer. If you play around with calculating S_n for various values of n this soon becomes easy to accept. It's also easy to see that once you start considering large integers, you can get a partial sum as close as you'd like to an integer value, because each new term adds so very little. For example, the first 272,400,599 terms give the partial sum 19.9999999979, but adding in just one more term pushes the partial sum past 20, to 20.0000000016. And that's what happens when you study *any* partial sum that is *just* less than an integer—the addition of one more term to the partial sum will *always overshoot* the integer value.

But doing specific, numerical studies isn't enough. A mathematician wants a *proof,* not cubic miles of printer paper. How can we prove this? It turns out to be easy, and here's how to do it.

Write N as the number that is the highest power of 2 that is not larger than n. That is,

$$N = 2^m \leq n,$$

where m is the largest integer such that this is so. Thus,

$$S_n = 1 + \frac{1}{2} + \frac{1}{3} + \frac{1}{4} + \frac{1}{5} + \frac{1}{6} + \frac{1}{7} + \frac{1}{8} + \ldots + \frac{1}{2^m} + \ldots + \frac{1}{n}.$$

Now, to write S_n as a fraction—with a finite number of terms, it is clear that S_n is a rational number given by the fraction p/q, because we can *calculate* p and q for any given n—we start by finding q. We do this by factoring each of the individual denominators $(1, 2, 3, \ldots, 2^m, \ldots, n)$ into a product of primes and selecting the highest power of each of the distinct factors that appear. We have already done that for the factor 2 (the denominator 2^m). So, all the remaining distinct factors must be *odd* integers, since 2 is the only even prime.

Powers of odd numbers are odd, and the product of odd numbers is odd, so the result of multiplying all these various powers together to form q (the so-called least common denominator) must be of the form "odd times2^m" which is, of course, even. Thus, if we write each term of S_n as a fraction with a denominator of q, the numerator of each such fraction will be even *except* for the $\frac{1}{2^m}$ term, which will have an odd numerator, because that numerator will have only odd numbers as factors. So,

$$S_n = \frac{\text{even}}{q} + \frac{\text{even}}{q} + \frac{\text{even}}{q} + \frac{\text{even}}{q} + \ldots + \frac{\text{odd}}{q} + \frac{\text{even}}{q} + \ldots + \frac{\text{even}}{q}$$

$$= \frac{\text{odd}}{q} = \frac{\text{odd}}{\text{even}},$$

since the sum of any number of evens and *one* odd is odd. But $\frac{\text{odd}}{\text{even}}$ cannot be an integer (!), and we are done.

Now, suppose we sum a series that looks a lot like the harmonic series except that an *infinite* number of the terms are missing. In particular, suppose we sum the reciprocals of *just the primes*. The big surprise, known since 1737, is that the sum *still* diverges to infinity:

$$\frac{1}{2} + \frac{1}{3} + \frac{1}{5} + \frac{1}{7} + \frac{1}{11} + \frac{1}{13} + \ldots = \infty.$$

This result, due to Euler, is another proof of the infinity of the primes (but Euclid's is still the better one). If we sum over only the *twin* primes, then, at last, the sum is *finite*:

$$B = \left(\frac{1}{3} + \frac{1}{5}\right) + \left(\frac{1}{5} + \frac{1}{7}\right) + \left(\frac{1}{11} + \frac{1}{13}\right) + \left(\frac{1}{17} + \frac{1}{19}\right) \ldots$$

$$= 1.90216 \ldots,$$

a value called *Brun's constant*, after the Norwegian mathematician Viggo Brun (1885–1978), who proved this result in 1919. This shows that, in a certain sense, the twin primes are "less dense" than are the primes (no surprise there), but it does *not* say anything one way or the other about the infinity (or not) of the twin primes. Still, despite the "small" value of B, nearly all mathematicians would, I think, still bet on the infinity of the twin primes.

Now, one more question concerning the primes: do the partial sums of the reciprocals of just the primes, alone, ever equal an integer? That is, is there some integer n such that

$$S_n = \frac{1}{p_1} + \frac{1}{p_2} + \frac{1}{p_3} + \ldots + \frac{1}{p_n}$$

equals an integer, where p_i is the ith prime? The answer is no, and this follows from the same argument that showed the partial sums of the harmonic series never equal an integer (because $p_1 = 2$ is the only even prime). As far as I know, there is no application either in mathematics or physics in which the possibility (or not) of an integer partial sum of either the harmonic series or the reciprocals of the primes plays a role. These demonstrations *are* nonetheless important, because they dramatically illustrate how mathematicians reason (which, while sometimes not quite how mathematical physicists reason, is somewhat similar).

The connection between pure math and the physical world clearly has an ancient origin (think of the subject of geometry, which literally means the use of logical reasoning to "measure the earth"). Another ancient math/physical world connection can be found in the association of irrational *numbers* with the geometric construction of line segments with irrational lengths. The concept of an irrational number (a number that cannot be written as $\frac{m}{n}$, that is, as a *ratio* of two integers) is usually introduced to students in high school; the demonstration that $\sqrt{2}$ is such a number, dating to the Greek mathematician Pythagoras (569–500 BC), is the classic example.

The beautiful argument attributed to Pythagoras, using the ideas of evenness and oddness, is so well presented in countless high school geometry books that I can't justify repeating it yet again here. But what I *will* show you is a less well known but equally beautiful analysis that has the virtue of showing *more* for even *less* effort. Specifically, I will show that the kth root ($k \geq 2$) of *any* prime (of which 2 is the first) is irrational and so, in addition to showing that $\sqrt{2}$ is irrational, we'll simultaneously conclude that $\sqrt{3}$, $\sqrt{5}$, $\sqrt{7}$, $\sqrt{11}$, and so on to infinity, are *all* irrational.[5] And the cube roots, the fourth roots, the fifth roots,

---

[5] The square root of *any* positive integer that isn't a perfect square is irrational, but limiting ourselves to the primes makes the proof a bit more straightforward.

and so on, of *any* prime, are irrational, too.[6] And best of all, while our proof will cover a lot of ground, it isn't any longer or more difficult to understand than is the Pythagorean one for the *lone* $\sqrt{2}$. So, here we go.

Assertion: $\sqrt[k]{p}$ is irrational for all $k \geq 2$ if p is a prime.

Proof: *Assume* that $\sqrt[k]{p}$ is rational. That is, *assume* that there are integers m and n so that $\sqrt[k]{p} = \frac{m}{n}$, $k \geq 2$. Thus, $p = \frac{m^k}{n^k}$. By the *unique* factorization of any positive integer into a product of primes (with some primes perhaps appearing multiple times), we can write

$$m = p_1 p_2 p_3 \ldots p_r, \quad n = q_1 q_2 q_3 \ldots q_s,$$

and so

$$p = \frac{(p_1 p_2 p_3 \ldots p_r)^k}{(q_1 q_2 q_3 \ldots q_s)^k},$$

or

$$p \, (q_1 q_2 q_3 \ldots q_s)^k = (p_1 p_2 p_3 \ldots p_r)^k = N,$$

where all the q_j's and p_i's are primes. Now—a *crucial* observation—the left- and right-hand sides of this equality represent the same number, N. This may strike you as so obvious as to be silly to even bother pointing out, but you'll soon see that it *is* crucial.

The starting p at the far left of the equality may in fact be one of the q_j or p_i, or a yet-different prime altogether. That is, on the left-hand side of the equality, none, one, or more than one of the q_j might be p. So, let's suppose that a of the q_j are p, where $a \geq 0$. Thus, on the left-hand side of the equality, p appears $ka + 1$ times.

On the right-hand side of the equality, none, one, or more than one of the p_i might be p. Let's suppose b of the p_i are p, where $b \geq 0$. Thus, on the right-hand side p appears kb times. Now, *unique* factorization of N into primes requires that $ka + 1 = kb$, which says that

$$b - a = \frac{1}{k}.$$

---

[6]While the irrationality of $\sqrt{2}$ is a standard high school math demonstration, I'm pretty sure that the irrationality of $\sqrt[3]{2}$ is not.

But obviously, there are *no* integers a and b such that their difference is less than 1 (which is what $\frac{1}{k}$ *is* for $k \geq 2$). So, the assumption of the rationality of $\sqrt[k]{p}$ has, *very quickly*, led us to an impossibility. Thus, $\sqrt[k]{p}$ is irrational, and our proof is done.

The ancient Greeks were fascinated by the irrational square roots of the nonsquare integers and, centuries before Christ, had discovered how to geometrically construct line segments of irrational length. In fact, given a line segment defined to be of unit length, they could then draw a line segment of length $\sqrt{n}$, where n is *any* positive integer (prime or otherwise). That can be done using what is called the *spiral of Theodorus*.[7]

All of Theodorus's actual work has been lost to history, and so it is not specifically known today just *how* he demonstrated the construction of the square roots of the integers from 3 to 17. The belief that he *did* so is based on the claim of the achievement included by Plato in his dialogue *Theaetetus*.[8] It has been suggested, however, that Theodorus's demonstration was, in some manner, based on the spiral of right triangles shown in Figure A1.1.

To make the spiral, a common vertex is used, and for each new triangle, the side opposite the common vertex has length 1. Since the other side that forms the right angle is the hypotenuse of the previous triangle, the length of the nth hypotenuse is $\sqrt{n+1}$. (The first such triangle has unit length for the two sides that form the right angle, and so a hypotenuse of $\sqrt{2}$.) Plato wondered why Theodorus stopped with the construction of $\sqrt{17}$ ("somehow he got into difficulties"), but Figure A1.1 offers us an obvious clue. If we compute the sum of the vertex angles for the first n triangles in the spiral, then we have for that sum

$$\sum_{k=1}^{n} \tan^{-1}\left(\frac{1}{\sqrt{k}}\right),$$

---

[7]After the Greek mathematician Theodorus of Cyrene (470–390 BC), who taught mathematics to Plato.

[8]Named after another student of Theodorus, who showed the irrationality of *all* the square roots of the nonsquare integers.

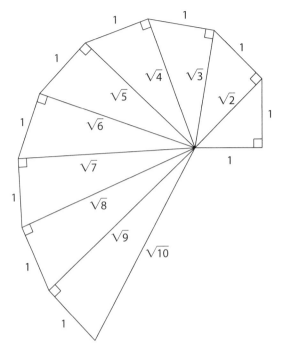

Figure A1.1. Theodorus's spiral of right triangles.

which for $n = 16$ is $351.15°$, while for $n = 17$ the sum is $364.78°$. That is, perhaps Theodorus stopped at $n = 16$ (which generates the triangle with hypotenuse $\sqrt{17}$) because, for $n > 16$, the spiral starts to overlap itself, and the construction becomes messy. Unless physicists someday give mathematicians a time machine, we'll probably never know the *real* story.

Now, here's a little challenge for you. What's the value of n when the spiral has overlapped itself for the *second* time? That is, when does the above sum first exceed 4π radians ($720°$)? You'll find a MATLAB solution at the end of this essay.

The moral of these last results is that while computers are useful, sometimes you really do have to *think* (not just program) your way to a solution. As a final example of using a computer and MATLAB in mathematical problems, consider the ancient problem of calculating the square root of a positive number—mundane, yes, perhaps, but undeniably highly useful to mathematical physicists.

Of course, MATLAB and similar languages have a built-in square-root command (in MATLAB, sqrt(n)), but I'll use MATLAB to show you a purely arithmetical way to do square roots by hand (recall my comment in the preface about being stranded on a sandy beach).

I remember in high school once reading in an algebra text a rule for hand-calculating square roots. It was a complicated sequence of operations, none of which I remember as having any obvious motivation, and I didn't understand it then. A far better way to illustrate square-root calculations is, I think, with what mathematicians call *continued fractions*, a subject that originated with the Italian engineer-architect Rafael Bombelli (1526–1572). While he didn't actually write the following expressions, they are implicit in his influential book *L'Algebra Opera* (1572):

$$\sqrt{2} = 1 + \cfrac{1}{2 + \cfrac{1}{2 + \frac{1}{2 + \ldots}}},$$

and

$$\sqrt{13} = 3 + \cfrac{4}{6 + \cfrac{4}{6 + \frac{4}{6 + \ldots}}},$$

where the ellipsis dots mean "and so on, forever."

If, for example, you numerically evaluate the $\sqrt{2}$ expression over and over, each time going successively "deeper" into the divisions, you get

$$\sqrt{2} \approx 1 + \frac{1}{2} = 1.5,$$

$$\sqrt{2} \approx 1 + \cfrac{1}{2 + \frac{1}{2}} = 1.4,$$

$$\sqrt{2} \approx 1 + \cfrac{1}{2 + \cfrac{1}{2 + \frac{1}{2}}} = 1.41666\ldots,$$

$$\sqrt{2} \approx 1 + \cfrac{1}{2 + \cfrac{1}{2 + \cfrac{1}{2 + \frac{1}{2}}}} = 1.413793\ldots,$$

and so on. These successive, so-called convergents, really do look as though they are converging to $\sqrt{2} = 1.414213\ldots$. (Try it for yourself with the $\sqrt{13}$ fraction; the only math involved is addition and division.) Both expressions are special cases of the general (and very pretty) formula

$$\sqrt{a^2 + b} = a + \cfrac{b}{2a + \cfrac{b}{2a + \cfrac{b}{2a + \cfrac{b}{2a + \cdots}}}}.$$

Here's how to derive this general expression[9] (we get Bombelli's $\sqrt{2}$ expression using $a = b = 1$, and the one for $\sqrt{13}$ using $a = 3$ and $b = 4$) in just a few easy lines of simple, first-year algebra. We start with

$$s^2 - a^2 = b,$$

where a and b are arbitrary positive numbers, and then we solve for s. That is,

$$s = \sqrt{a^2 + b}.$$

Now, factoring the original equation, we get

$$(s + a)(s - a) = b,$$

or

$$s - a = \frac{b}{s + a},$$

and so,

$$s = a + \frac{b}{s + a} = a + \frac{b}{a + s}.$$

---

[9]A few decades later (1613) Bombelli's fellow Italian Pietro Cataldi (1548–1626) showed he may have known this, too, when he wrote the continued fraction for $\sqrt{18}$ with $a = 4$ and $b = 2$.

If we now continuously substitute for s on the right with what this last result says s equals, we have

$$s = a + \cfrac{b}{a + a + \cfrac{b}{a + a + \cfrac{b}{a+a+\frac{b}{a+a+\dots,}}}}$$

and, just like that, we have our result:

$$\sqrt{a^2 + b} = a + \cfrac{b}{2a + \cfrac{b}{2a+\cfrac{b}{2a+\frac{b}{2a+\dots}}}}.$$

Now, for a specific illustrative example from history, consider the quadratic equation

$$x^2 - x - 1 = 0.$$

We know, from the quadratic formula, that

$$x = \frac{1 \pm \sqrt{1+4}}{2} = \frac{1 \pm \sqrt{5}}{2}.$$

The positive root is, in particular, a number famous in mathematics as the *golden mean*,

$$x = 1.61803398874989\dots,$$

which the ancient Greeks thought to be the ratio of the length to the width of the most aesthetically pleasing rectangle. To calculate this value as a continued fraction, we write the original quadratic as

$$x(x-1) = 1,$$

and so

$$x - 1 = \frac{1}{x},$$

or

$$x = 1 + \frac{1}{x}.$$

Then, using the idea of continuously replacing x on the right with the expression for x in this last equation, we get the continued fraction

$$x = 1 + \cfrac{1}{1 + \cfrac{1}{1 + \cfrac{1}{1 + \cfrac{1}{1 + \cdots}}}}.$$

Let's call x_n the value we get if we truncate this continued fraction by using just the first n division bars; that is (for example),

$$x_1 = 1 + \frac{1}{1} = 2 \quad \text{and} \quad x_2 = 1 + \cfrac{1}{1 + \frac{1}{1}} = 1 + \frac{1}{2} = 1.5.$$

The calculation of x_n by hand becomes ever more tedious as n increases, and so I'll now show you how to use MATLAB to do all the arithmetic. In particular, we'll answer these two questions: (a) what's the value of x_{17}? and (b) what's the value of the first n so that x_n has the first nine digits correct in the code's estimate for the golden mean? Both questions would be horrible to attempt by hand, but MATLAB removes the pain. In the code **gm.m** the calculations are done from the "bottom up":

```
%gm.m/created by PJNahin for Mathematical Physics
n=input('How many division bars?')
x=2;
while n>1
    x=1+1/x;
    n=n-1;
end
x
```

Running **gm.m** gives us (in a flash) that $x_{17} = 1.618034055\ldots$, and a few additional trial-and-error runs tell us that $x_{21} = 1.618033990\ldots$ (which has only the first seven digits correct), while $x_{22} = 1.618033988\ldots$ which has the first nine digits correct. So, the answer to (b) is $n = 22$.

A more spectacular example is the continued fraction for $\frac{4}{\pi}$, found by the Englishman William Brouncker (1620–1684) in 1656. It is considerably more complicated to derive than that for the golden mean fraction,[10] and it converges *much* more slowly:

$$\frac{4}{\pi} = 1 + \cfrac{1^2}{2 + \cfrac{3^2}{2 + \cfrac{5^2}{2 + \cfrac{7^2}{2 + \cdots}}}}.$$

The code **fouroverpi.m** calculates the convergents of this continued fraction, using just the first n division bars, and as with **gm.m**, the calculations are done from the bottom up.

```
%fouroverpi.m/created by PJNahin for Mathematical Physics
n=input('How many division bars?')
x=2;
while n>1
    k=2*n - 1;
    x=2 + (k^2)/x;
    n=n - 1;
end
x=1+1/x
```

The value of $\frac{4}{\pi}$ is $1.2732395447\ldots$, while using 200 division bars (not something *you'd* want to do!) in the continued fraction gives a value of $1.2712264\ldots$ with only the first two digits correct. Using 2,000 (!) division bars gives only one more correct digit ($1.27303\ldots$).

Brouncker's continued fraction shows that while the digits of π seem to display all the characteristics generally associated with randomness (the most obvious being the equal probability of occurrence of every digit from 0 to 9), there nonetheless must be a decided *structure* to π, too. The nature of that structure has a very deep feeling of mystery to it, however, the least of which is the question, just what causes π to appear in so many of the equations encountered by mathematical physicists when they study the physical world?

---

[10] Because I know you're certain to be curious, you'll find a derivation in Appendix 2.

Okay, have you had success in solving the challenge problem I gave you earlier, concerning the Diophantine equation $y^2 - 17 = x^3$? Here's how I did it. The following MATLAB code, **equation.m**, ran on my laptop in less than one second, generating all the integer values for x that give integer values for y. The results were the values I gave you earlier, plus the next much larger solution, $x = 5{,}234$. There are no other positive integer solutions.

```
%equation.m/created by PJNahin for Mathematical Physics
for x=1 : 10000
    cube=x ^ 3;
    y=sqrt(cube + 17);
    if y==round(y)
        x
    end
end
```

The code **theo.m** shows, in answer to the challenge question I gave you earlier, that the second overlap in Theodorus's spiral starts with $n = 54$ (the total vertex sum is $727.48°$).

```
%theo.m/created by PJNahin for Mathematical Physics
c=4*pi; s=0; k=0;
while s < c
    k = k+1;
    s = s+atan(1/sqrt(k));
end
k, (180/pi)*s
```

A Final Comment

In the preface, and in this appendix, I've briefly discussed the use of computers in mathematical physics work. My remarks, and the computer codes included in this book, are all in the traditional sense as commonly viewed today. For a radically different view, by one of the more famous mathematical physicists of the 20th century, see Richard P. Feynman, "Simulating Physics with Computers," *International Journal of Theoretical Physics*, June 1982, pp. 467–488.

Appendix 2
A Derivation of Brouncker's Continued Fraction for $\frac{4}{\pi}$

The analysis that follows is not how Brouncker did it but instead
has been intentionally crafted so as to be well within the reach of
a good high school AP Calculus student.

Suppose we write the following continued fraction with an obvious pattern:

$$N = c_0 + \cfrac{c_1}{1 - \cfrac{c_2}{1+c_2 - \cfrac{c_3}{1+c_3 - \cfrac{c_4}{1+c_4 - \cdots}}}}.$$

If we evaluate the first four convergents, we get (where the subscript on N is the number of division bars used)

$N_0 = c_0,$

$N_1 = c_0 + c_1,$

$N_2 = c_0 + \cfrac{c_1}{1 - \cfrac{c_2}{1+c_2}} = c_0 + \cfrac{c_1}{\frac{1+c_2-c_2}{1+c_2}} = c_0 + c_1(1+c_2) = c_0 + c_1 + c_1 c_2,$

and

$$N_3 = c_0 + \cfrac{c_1}{1 - \cfrac{c_2}{1+c_2 - \frac{c_3}{1+c_3}}} = c_0 + c_1 + c_1 c_2 + c_1 c_2 c_3,$$

where I'll let you fill in the simple algebraic steps for the last calculation. And so we see what I think is a pretty obvious pattern in the successive convergents that suggests the value of the continued fraction is given by the infinite series

$$\lim_{n \to \infty} N_n = N = c_0 + c_1 + c_1 c_2 + c_1 c_2 c_3 + c_1 c_2 c_3 c_4 + \cdots .$$

Next, consider (with, I admit, absolutely no motivation!) the power series expansion for $\tan^{-1}(x)$. We start with the differentiation formula[1]

$$\frac{d}{dx} \tan^{-1}(x) = \frac{1}{1+x^2},$$

and so, performing the long division on the right, we have

$$\frac{d}{dx} \tan^{-1}(x) = 1 - x^2 + x^4 + \cdots + (-1)^n x^{2n} + \cdots .$$

Then, integrating indefinitely, we get

$$\tan^{-1}(x) + C = x - \frac{1}{3}x^3 + \frac{1}{5}x^5 + \cdots + (-1)^n \frac{1}{2n+1} x^{2n+1} + \cdots ,$$

where C is an arbitrary constant. But, since $\tan^{-1}(0) = 0$, then $C = 0$, and so

$$\tan^{-1}(x) = x - \frac{1}{3}x^3 + \frac{1}{5}x^5 + \cdots + (-1)^n \frac{1}{2n+1} x^{2n+1} + \cdots .$$

Returning to

$$N = c_0 + c_1 + c_1 c_2 + c_1 c_2 c_3 + c_1 c_2 c_3 c_4 + \cdots ,$$

---

[1]This differentiation formula is just the inverse of the freshman calculus indefinite integration formula $\int \frac{dx}{1+x^2} = \tan^{-1}(x) + C$, where C is an arbitrary constant. This integration formula is one all mathematical physicists know (and should know how to *derive* with a stick on a sandy beach). Can you do that? If not, see the end of this appendix. Differentiating both sides of the integration formula immediately gives us $\frac{d}{dx} \tan^{-1}(x) = \frac{1}{1+x^2}$.

we set (you'll see why, soon)

$$c_0 = 0,$$

$$c_1 = x,$$

$$c_1 c_2 = x c_2 = -\frac{1}{3} x^3, \quad \text{or} \quad c_2 = -\frac{1}{3} x^2,$$

$$c_1 c_2 c_3 = x \left(-\frac{1}{3} x^2 \right) c_3 = \frac{1}{5} x^5, \quad \text{or} \quad c_3 = -\frac{3}{5} x^2,$$

$$c_1 c_2 c_3 c_4 = x \left(-\frac{1}{3} x^2 \right) \left(-\frac{3}{5} x^2 \right) c_4 = -\frac{1}{7} x^7, \quad \text{or} \quad c_4 = -\frac{5}{7} x^2,$$

and so on. What we've done with these choices for the c's, of course, is purposely to have *forced* $N = \tan^{-1}(x)$. So, substituting these values for the c's into our continued fraction, we have

$$N = \cfrac{x}{1 - \cfrac{-\frac{1}{3}x^2}{1 - \frac{1}{3}x^2 - \cfrac{-\frac{3}{5}x^2}{1 - \frac{3}{5}x^2 - \cfrac{-\frac{5}{7}x^2}{1 - \frac{5}{7}x^2 - \cdots}}}}$$

$$= \cfrac{x}{1 + \cfrac{\frac{1}{3}x^2}{\frac{2}{3}x^2 + \cfrac{\frac{3}{5}x^2}{\frac{2}{5}x^2 + \cfrac{\frac{5}{7}x^2}{\frac{2}{7}x^2 + \cdots}}}} = \cfrac{x}{1 + \cfrac{x^2}{2x^2 + \cfrac{\frac{3^2}{5}x^2}{\frac{2}{5}x^2 + \cfrac{\frac{5^2}{7}x^2}{\frac{2}{7}x^2 + \cdots}}}}$$

$$= \cfrac{x}{1 + \cfrac{x^2}{2x^2 + \cfrac{3^2 x^2}{2x^2 + \cfrac{\frac{5^2}{7}x^2}{2x^2 + \cdots}}}} = \cfrac{x}{1 + \cfrac{x^2}{2x^2 + \cfrac{3^2 x^2}{2x^2 + \cfrac{5^2 x^2}{2x^2 + \cdots}}}}$$

or, setting $x = 1$, we have $N = \tan^{-1}(1) = \frac{\pi}{4}$, and so

$$\frac{\pi}{4} = \cfrac{1}{1 + \cfrac{1^2}{2 + \cfrac{3^2}{2 + \cfrac{5^2}{2 + \cfrac{7^2}{2 + \cdots}}}}}$$

or, inverting both sides, we have our result:

$$\frac{4}{\pi} = 1 + \cfrac{1^2}{2 + \cfrac{3^2}{2 + \cfrac{5^2}{2 + \cfrac{7^2}{2 + \cdots}}}}.$$

Deriving $\int \frac{dx}{1+x^2} = \tan^{-1}(x)$ with a stick on a sandy beach, we draw the right triangle

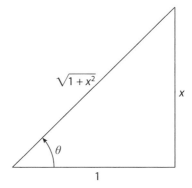

which says that $x = \tan(\theta) = \frac{\sin(\theta)}{\cos(\theta)}$. That is, $\theta = \tan^{-1}(x)$. Thus, $\frac{dx}{d\theta} = \frac{\cos^2(\theta)+\sin^2(\theta)}{\cos^2(\theta)} = \frac{1}{\cos^2(\theta)}$, or $dx = \frac{d\theta}{\cos^2(\theta)}$. Since $\cos(\theta) = \frac{1}{\sqrt{1+x^2}}$, then $\frac{1}{\cos^2(\theta)} = 1+x^2$, and so $dx = (1+x^2)d\theta$. Thus, $\int \frac{dx}{1+x^2} = \int \frac{(1+x^2)d\theta}{1+x^2} = \int d\theta = \theta = \tan^{-1}(x)$. Done.

Appendix 3
Landen's Calculus Solution to the Depressed Cubic Equation

This essay is a curious story about a very smart fellow, John Landen (1719–1790), who while known to a few of today's mathematicians in a highly specialized area of integral theory[1] has otherwise effectively vanished from the history of mathematics. Even though self-taught, Landen was eventually elected a Fellow of the Royal Society in recognition of the high regard in which he was held by academic mathematicians of the day. He did a number of things to earn that regard, one of which was to publish a beautiful, extremely clever way to solve cubic equations.[2]

If you're up on your math history, the date (1755) of Landen's discovery might surprise you, since it had been known since the beginning of the 16th century how to solve what is called the *depressed cubic* $x^3 + px = q$ (called *depressed* because the second-degree term

---

[1]Specifically, the theory of elliptic integrals, which we will *not* get into in this book (but see my book *In Praise of Simple Physics*, Princeton University Press, 2016, pp. 163–176, for an example of an elliptic integral in a modern mathematical physics problem). A very nice account of just what Landen did in that area was the basis for the 1933 Presidential Address to the London Mathematical Society: see G. N. Watson, "The Marquis and the Land-Agent: A Tale of the Eighteenth Century," *The Mathematical Gazette*, February 1933, pp. 5–17. As you continue to read this essay, the significance of Watson's title will become clear.

[2]In his 1755 book *Mathematical Lucubrations*. This odd title was chosen by Landen to alert his readers to the fact that it was a most serious work, the result of sustained, intense effort.

is missing), and that allows you to solve the general cubic with *all* the terms present.[3] That is, what Landen did in 1755 had been done more than 200 years earlier![4] So, you ask, what's the big deal?

The big deal is that all that earlier work had been algebraic in nature, as the calculus hadn't yet been invented by either Newton in England or Leibniz in Germany (dating to, for Newton, around 1670 or so). Landen showed how to use the "new" tool of calculus to solve cubics. Nearly two hundred years after Landen's death his method attracted the attention of a physicist who noted that a well-known modern book on advanced college algebra simply gave Landen's solution, without either a derivation or even a citation.[5] As Problem P19 in this book shows, cubic equations do occur in physics, applied mathematics, and engineering, and knowing how to solve them is an essential technical skill expected of a mathematical physicist.

I suspect most (if not all) physicists have learned how to solve cubics via the algebraic route, and Landen's calculus approach will be an eye-opener. That's because while algebra *does* do the job, it is *just* up to doing it, and so those early derivations are extremely clever and tricky. Landen, however, had both the differential and integral calculus available to him by 1755, and his solution to the cubic equation is essentially routine (or at least it is after you see it). The algebraic solutions, however, look astonishingly clever, almost magical, even *after* you've read through them. (That's my personal opinion, of course, and many mathematicians may disagree with me—but I think they are just fooling themselves!)

---

[3]In the general cubic equation $y^3 + ay^2 + by + c = 0$, change the variable to $y = x - \frac{1}{3}a$. After substitution, expansion, and collection of terms, you'll find that the x^2 term has vanished, leaving just a depressed cubic in x, with $p = b - \frac{1}{3}a^2$, and $q = \frac{1}{3}ab - \frac{2}{27}a^3 - c$.

[4]You can read all about that early work on the cubic in my book *An Imaginary Tale: The Story of $\sqrt{-1}$*, (in several editions), Princeton University Press, pp. 8–30.

[5]William Squire, "Landen's Solution of the Cubic," *American Journal of Physics*, April 1987, pp. 374–375 (with a correction published in December 1987, p. 1146). The "well-known" math book was Birkhoff and McLane's *A Survey of Modern Algebra*, Macmillan, 1953, p. 96.

To give you the flavor of Landen's method, here's how it works on a special case of the depressed cubic. Starting with

$$x^3 + px = q = x(x^2 + p), \tag{1}$$

where p and q are each arbitrary *positive* constants, Landen generalized the nature of q (but not that of p) to be a function of x. That is, $q = q(x)$. This, of course, includes the possibility that q is a constant, as a special case. Then, differentiating (1) twice with respect to x, we have

$$\dot{q} = 3x^2 + p, \tag{2}$$

and

$$\ddot{q} = 6x, \tag{3}$$

where I've used (as did Landen) Newton's single- and double-dot notation to represent the first and second derivatives of q, respectively. The dot notation appears, from time to time, in physics journals to this day, usually as a time derivative, but in Landen's work the differentiation is done with respect to x (which might, in fact, be time, as in Problem P18).

If we set $x = 0$, then these three equations give us the conditions

$$q(0) = 0, \quad \dot{q}(0) = p, \quad \ddot{q}(0) = 0, \tag{4}$$

which will be very useful to us as we get into the analysis. From (1) we have

$$x = \frac{q}{x^2 + p},$$

and putting this into (3), we get

$$\ddot{q} = \frac{6q}{x^2 + p}. \tag{5}$$

From (2) we have

$$x^2 = \frac{\dot{q} - p}{3},$$

and putting this into (5) gives us

$$\ddot{q} = \frac{6q}{\frac{\dot{q}-p}{3}+p} = \frac{18q}{\dot{q}+2p},$$

and so, cross multiplying, we obtain

$$\ddot{q}\dot{q} + 2p\ddot{q} = 18q. \qquad (6)$$

Multiplying through (6) with $\dot{q}$ gives us

$$\ddot{q}\dot{q}^2 + 2p\dot{q}\ddot{q} = 18q\dot{q}, \qquad (7)$$

which integrates (by inspection) to

$$\frac{1}{3}\dot{q}^3 + p\dot{q}^2 = 9q^2 + C_1, \qquad (8)$$

where C_1 is some constant. If (8) isn't clear, then simply differentiate (8) and verify that the result is indeed (7). The value of C_1 is easily found by using the conditions given in (4). That is, for $x = 0$ we have

$$\frac{1}{3}p^3 + p^3 = C_1 = \frac{4}{3}p^3,$$

and so (8) becomes

$$\frac{1}{3}\dot{q}^3 + p\dot{q}^2 = 9q^2 + \frac{4}{3}p^3,$$

or

$$\dot{q}^3 + 3p\dot{q}^2 = 27q^2 + 4p^3 = \dot{q}^2(\dot{q} + 3p). \qquad (9)$$

Since we know from (2) that

$$\dot{q} = 3x^2 + p,$$

then (9) says

$$\dot{q}^2\left(3x^2 + 4p\right) = 27q^2 + 4p^3 = \left(\frac{dq}{dx}\right)^2 (3x^2 + 4p),$$

or

$$\frac{dq}{dx} = \frac{\sqrt{27q^2 + 4p^3}}{\sqrt{3x^2 + 4p}},$$

which allows us to separate x and q and so write

$$\frac{dq}{\sqrt{27q^2 + 4p^3}} = \frac{dx}{\sqrt{3x^2 + 4p}}.$$

Factoring 27 out of the denominator on the left and 3 out of the denominator on the right gives us

$$\frac{dq}{\sqrt{q^2 + \frac{4}{27}p^3}} = 3 \frac{dx}{\sqrt{x^2 + \frac{4}{3}p}}. \tag{10}$$

We find the following entry in integral tables:

$$\int \frac{dy}{\sqrt{y^2 + a^2}} = \sinh^{-1}\left(\frac{y}{a}\right).$$

This is exactly the form[6] we have on both sides of (10) because of our assumption that $p > 0$. So, integrated indefinitely, (10) tells us that

$$\sinh^{-1}\left(\frac{q}{\sqrt{\frac{4}{27}p^3}}\right) + C_2 = 3\sinh^{-1}\left(\frac{x}{\sqrt{\frac{4}{3}p}}\right).$$

When $x = 0$, we know from (4) that $q = 0$, and so, since $\sinh^{-1}(0) = 0$, we have $C_2 = 0$. Thus,

$$x = 2\sqrt{\frac{p}{3}}\sinh\left\{\frac{1}{3}\sinh^{-1}\left(\frac{3}{2}q\sqrt{\frac{3}{p^3}}\right)\right\}, \quad p, q > 0. \tag{11}$$

---

[6]At the risk of stating the obvious, the *hyperbolic sine* is defined as $\sinh(x) = \frac{e^x - e^{-x}}{2}$, and the *inverse hyperbolic sine* is written as $\sinh^{-1}(x)$. That is, $x = \sinh^{-1}\{\sinh(x)\}$.

Let's check a couple of specific cases of (11) to see if it works. First, suppose we have $x^3 + 6x = 20$; that is, $p = 6$ and $q = 20$. You can quickly see, *by inspection*, that $x = 2$ works.[7] The result in (11) agrees because

$$x = 2\sqrt{\frac{6}{3}}\sinh\left\{\frac{1}{3}\sinh^{-1}\left(30\sqrt{\frac{3}{216}}\right)\right\},$$

which, hand calculated, does indeed give $x = 2$. So, (11) has correctly found the one real root to $x^3 + 6x = 20$. Now all cubics have *three* roots, of course, so what about the other two? They are easily found by dividing $x^3 + 6x - 20$ by $(x - 2)$, to arrive at a quadratic equation, which we already know how to solve. Here's another example, where the root is not obvious by inspection: let's find the real, positive root to $x^3 + 9x - 2 = 0$. We have $p = 9$ and $q = 2$, and so (11) says

$$x = 2\sqrt{\frac{9}{3}}\sinh\left\{\frac{1}{3}\sinh^{-1}\left(\frac{3}{2}2\sqrt{\frac{3}{729}}\right)\right\} = 0.22102253\ldots,$$

which is easily confirmed to satisfy the cubic. For the cases where $p > 0$ and/or $q > 0$ are *not* true, I'll simply refer you to Squire's paper (note 5) for additional details.

Who was the man who did this elegant work? John Landen was born a few miles from Peterborough, England, and, almost to the day, died 71 years later not very far from his birthplace. After no more than the typical education for a boy of common birth—the village school at the local church, followed (perhaps) with some finishing touches at an "advanced" grammar school—he took up the work of a surveyor. In that way he earned his living by day, from 1740 to 1762.

In the evenings, however, his mind turned to mathematical puzzles far beyond a surveyor's simple trigonometry and geometry, and he started submitting what turned out to be a long series of increasingly

---

[7]When p and q are both positive (as in our specific case), there is only one real, positive root to the depressed cubic. A geometric proof of this is given on p. 11 of *An Imaginary Tale* (note 4). For the $p < 0$ case the situation is just a bit more complicated. Then, the possibility exists for either one real root *or* for three real roots. (It is impossible for there to be either no real roots or two real roots, as complex roots always appear as *pairs of conjugates*.)

deeper contributions to the *Ladies' Diary*. Despite its name, this was a respected outlet for mathematicians, both amateur and professional, in which to publish.[8] It wasn't nearly as stuffy as some modern math journals are, and the use of funny pseudonyms was not discouraged: it is known, for example, that Landen used such monikers as *C. Bumpkin*, *Peter Puzzlem*, and *Sir Stately Stiff*. From the *Ladies' Diary* he graduated (under his own name!) to the pages of the *Philosophical Transactions* of the Royal Society.[9] As his work continued to display an increasing level of sophistication Landen came to the favorable attention of both the mathematicians of the Royal Society (who elected him a Fellow in 1766), and those in France (where he was dubbed the "English D'Alembert"[10]).

In 1762 he became the land agent for a local member of the nobility, William Wentworth, Earl Fitzwilliam (who was, at one point, the second richest person in England), a position he held until 1788, two years before his death. During those years he continued his evening mathematical work, writing equations by candlelight. That work became progressively more difficult for him, however, as in his final years Landen suffered terribly from bladder stones, a debilitating, extremely painful affliction that kept him bedridden for weeks at a time. His death was probably a blessed relief from unending misery.

The final word that we have on Landen appeared more than 40 years after his passing, in *The Georgian Era*, a four-volume collection of biographical sketches that are, in words on the title page, "Memoirs of the Most Eminent Persons Who Have Flourished in Great Britain, from the Accession of George the First to the Demise of George the Fourth." Landen appears in volume 3, published in London in 1834,

---

[8]The *Ladies' Diary* was an annual publication that appeared from 1704 to 1841. With a circulation of several thousand, it offered a way for Landen to get his work out and known to like-minded math enthusiasts. You can find much more about this early math journal in Teri Perl, "The *Ladies' Diary* or *Woman's Almanack*, 1704–1841," *Historia Mathematica*, February 1979, pp. 36–53.

[9]A listing of Landen's publications can be found in H. Gwynedd Green and H.J.J. Winter, "John Landen, F.R.S. (1719–1790)—Mathematician," *Isis*, 35 (no. 1) 1944, pp. 6–10.

[10]The reference is to the French mathematical physicist Jean Le Rond d'Alembert (1717–1783).

devoted to "Voyagers and Travellers; Philosophers and Men of Science; Authors." There we read:

> Mr. Landen, undoubtedly, ranks very high as a mathematician; but his character appears to have been, in some respects, far from amiable. He possessed a coarseness of mind, which ... made him treat his inferiors with contempt. ... From the contrast between his manners, and those of his noble friend, the Earl of Fitzwilliam, the villagers are said to have been in the habit of exclaiming, when they saw them pass together, "There goes Lord Landen and Mr. Fitzwilliam." It is a fact, that his manuscripts were sold for waste paper, to the shopkeepers of Peterborough.

The person who wrote that summary obviously had a very harsh opinion of Landen's social skills (as a land agent, Landen surely had rubbed a lot of people the wrong way over the years): Landen's ghost must have been particularly outraged by that last sentence, one guaranteed to irritate *any* mathematician. But still, I think his ghost gets the last laugh. Nobody (or, at least, hardly anybody) remembers *The Georgian Era* today, but the next time you are faced with solving a cubic, you'd do well to remember Mr. Landen.

Finally, for those readers who may still think that Landen's use of calculus was, perhaps, a bit of overkill, let me argue otherwise. We can say a lot more about cubic equations with calculus than we can with just algebra. For example, in a previous book I presented (as stated in note 7) a simple, calculus-based demonstration that the depressed cubic equation $x^3 + px = q$, where p and q are both nonnegative, has exactly one real, *positive* solution.[11] In a later book, I asserted (in the midst of solving a problem involving Archimedes' principle) that the depressed cubic equation $r^3 - pr + q = 0$, where p and q are both positive, has exactly one real, *negative* solution.[12] I didn't prove that claim, but again an easy demonstration of it is calculus based, as follows.

We start with $f(r) = r^3 - pr + q$, $p, q > 0$ and wish to show that there is one real, *negative* solution to $f(r) = 0$. The other two solutions are either both real and *positive* or form a complex conjugate pair. Taking

---

[11] P. J. Nahin, *An Imaginary Tale*, Princeton University Press, 2010, p. 11.
[12] P. J. Nahin, *In Praise of Simple Physics*, Princeton University Press, 2016, p. 139.

the derivative, we have

$$f'(r) = 3r^2 - p,$$

and so

$$f'(r) = 0 \text{ at } r = \pm\sqrt{\frac{p}{3}}.$$

This tells us where a plot of $f(r)$ versus r has local extrema. Now, since the second derivative is

$$f''(r) = 6r,$$

we see that

$$f''\left(-\sqrt{\frac{p}{3}}\right) < 0,$$

while

$$f''\left(\sqrt{\frac{p}{3}}\right) > 0,$$

and these two results tell us that the extremum at $r = -\sqrt{\frac{p}{3}}$ is a local *maximum*, and the extremum at $r = \sqrt{\frac{p}{3}}$ is a local *minimum*. Also,

$$f(0) = q > 0.$$

Now, when $r \gg 0$, we have

$$f'(r) > 0, \quad f(r) > 0,$$

and when $r \ll 0$, we have

$$f'(r) > 0, \quad f(r) < 0.$$

These last two boxes tell us that the $f(r)$ versus r plot has a positive slope for (sufficiently) negative r *and* for (sufficiently) positive r, while

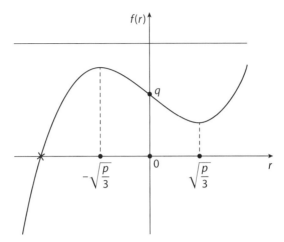

Figure A3.1. The cubic equation $r^3 - pr + q = 0$ with one real, *negative* solution.

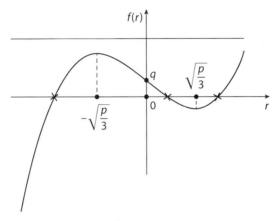

Figure A3.2. The cubic equation $r^3 - pr + q = 0$ with three real (one *negative*) solutions.

$f(r)$ itself is negative for (sufficiently) negative r and positive for (sufficiently) positive r. There are only two possible generic behaviors for $f(r)$ versus r that satisfy *all* these requirements, as shown in Figures A3.1 and A3.2.

The common feature of the two possibilities is that one solution is real and *negative*. For both, notice that $f\left(-\sqrt{\frac{p}{3}}\right) > q$ is required. This is easy to verify algebraically (try it!).

Appendix 4

Solution to Lord Rayleigh's Rotating-Ring Problem of 1876[1]

During the mid-Victorian age, the rotating ring would definitely have been considered a challenging exam problem in mathematical physics, just as Reverend Ward remembered it from his 1876 Mathematical Tripos examination. Today, however, it would be downgraded to being simply a good homework problem in a sophomore mechanical engineering class on the strength of materials. So, what happened in the century following that famous exam to account for such a dramatic change?

A modern student quickly "sees" the ultimate fate of the ring as its rate of rotation increases: each portion of the ring experiences an outward force (the *centrifugal force*, called a *fictitious force* by physicists, but a seemingly quite real force to all who have felt it while on a merry-go-round) that increases with the spin rate until the ring eventually explodes. In particular, an arbitrary half of the ring—see Figure A4.1— experiences a force that is attempting to separate it from the other half (which is "enjoying" the same experience).

And, interestingly, many of the 1876 math students saw that, too. We know this because some of their examination papers have survived;

---

[1] Besides Andrew Warwick's book (cited in the book epigraph), you can find Rayleigh's ring problem on page 280 of volume 1 of *The Scientific Papers of Lord Rayleigh*, Cambridge University Press, 1899. Alas, Rayleigh left us no solution, but perhaps the analysis I give here is something like what he had in mind. The references in this essay to "Poynting" are to John Poynting (mentioned in the footnote to the epigraph), the third-place finisher in the 1876 exam.

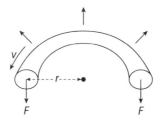

Figure A4.1. Half of Rayleigh's rotating ring.

indeed, Warwick reproduces Poynting's Tripos exam sheets, which show that he did in fact derive the correct *symbolic* expression for the stress in the rotating ring.[2] But what those exam sheets also show (on a point that Warwick passes over) is that nowhere does Poynting actually *calculate* anything. Even Rayleigh's numerical data were not reproduced, not even as a marginal doodle.

I believe that this absence of numerical calculation was because Poynting *simply didn't know how* to connect his *symbolic* solution with Rayleigh's numbers. "*How*," he must have frantically wondered to himself, with not just a bit of irony as the exam room clock relentlessly ticked away the seconds (the students were under intense time pressure, with just eighteen minutes available for each question), "*do I get a velocity (length per unit time) from 90,000 pounds per square inch and 490 pounds per cubic foot? How does time come in?*"

This is a fundamental issue that today's physics and engineering student is trained to always keep in mind: when deriving *symbolic* equalities, the expressions on each side of the equal sign *must* have the same units. Now this may seem so obvious as to be borderline silly to bother writing down, but in fact it is the basis for the amazingly powerful technique of *dimensional analysis*, a subject just beginning to be appreciated when Poynting sat for the 1876 Mathematical Tripos.[3] Less than a century later (1950), a dramatic illustration of its power

---

[2]Warwick correctly observes that these surviving examination scripts are "the closest thing we have to real-time records of mathematical physics in the making," as opposed to reading polished, published manuscripts. This feature of Warwick's book makes it both valuable for the historian as well as entertaining.

[3]Dimensional analysis, as a subject, is usually dated to an 1863 paper by the great Scottish mathematical physicist James Clerk Maxwell.

came when the English mathematical physicist Sir Geoffrey Taylor (1886–1975) used dimensional analysis to correctly derive the energy of the first American atomic bomb test of July 16, 1945 (code named "Trinity"). Working from nothing more than a widely published series of high-speed, time-lapse photographs of the explosion fireball that had been declassified in 1947, his calculation was so astonishingly accurate that U.S. authorities (who had tried to keep the explosion energy top secret) initially thought Taylor must have somehow breached military security.[4]

Let me now show you how an appreciation for the importance of units reduces Rayleigh's rotating-ring question to an undergraduate engineering homework problem. Imagine the ring, with radius r, rotating in a vertical plane: the half-ring is now as shown in Figure A4.2, where dm denotes a differential mass of the ring. The figure shows two such masses, one at angle θ and a matching mass at angle $180° - \theta$. The radially outward centrifugal force dF on each of the dm's can be resolved into horizontal and vertical components, where it is clear by symmetry that the two horizontal components cancel. The equal vertical components, however, add, and so if we sum (integrate) the vertical components over $0 \le \theta \le \pi/2$ radians, all we need do is double the result to get the total *vertical* centrifugal force acting on the half-ring.

The differential length of each differential mass is $dl = rd\theta$, and so, with a uniform cross-sectional area of a, the volume of a differential mass is adl. If the ring has density ρ, then the differential mass is $dm = adl\rho = a\rho rd\theta$. If the tangential speed of the ring is v, then the radially outward differential centrifugal force on a differential mass is, from the well-known formula for centrifugal force, given by

$$dF = \frac{v^2 dm}{r} = a\rho v^2 d\theta,$$

and so the *vertical, upward* component of the differential force on dm is given by

$$dF \sin(\theta) = a\rho v^2 \sin(\theta) d\theta.$$

---

[4]Taylor's dimensional analysis of the *Trinity* explosion can be understood by a high school student, and you can find a complete discussion of what he did in my book *In Praise of Simple Physics*, Princeton University Press, 2016, pp. 232–236.

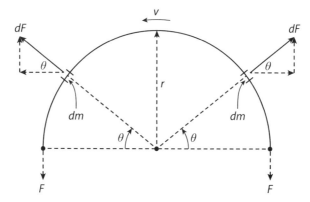

Figure A4.2. Two symmetrically located differential masses in the half-ring.

Thus, the total integrated vertical force on the half-ring is

$$2 \int_0^{\frac{\pi}{2}} a\rho v^2 \sin(\theta)\, d\theta = 2a\rho v^2 \left[-\cos(\theta)\right]\Big|_0^{\frac{\pi}{2}} = 2a\rho v^2.$$

This is the force that is balanced—until the ring comes apart—by the tensions in the two ends of the half-ring, so

$$2F = 2a\rho v^2,$$

or

$$F = a\rho v^2.$$

Assuming the wire's cross-sectional dimensions are "small" compared with r, the stress in the ring—at *every* point in the ring, not just at the ends of the imagined half-ring—is given by

$$\sigma = \frac{F}{a} = \rho v^2.$$

Thus,

$$v = \sqrt{\frac{\sigma}{\rho}}.$$

With this result, we see the independence of v on a (the wire's cross-sectional area) and on r (the radius of the ring), just as Rayleigh claimed. Both have canceled away.

Notice that v should have the units of speed (meters per second), and so this is a good place to stop for a moment to check units. When Rayleigh put this problem on the 1876 Mathematical Tripos, the metric system was just coming into place in England, where it was *not* immediately embraced with enthusiasm, and that's why we see him still using units like inches, cubic feet, and pounds. So, before going any further, let's convert his numbers and units into the modern MKS system, where we measure mass, length, time, energy, and force in units of kilograms, meters, seconds, joules, and newtons, respectively.

To start, if we put a 1-kilogram mass on a scale that reads in pounds, we'll see the pointer indicate 2.2 pounds. Also, 1 meter equals 39.37 inches, or 3.28 feet. From these two facts, we can convert Rayleigh's numbers/units to MKS. First, a density for steel of 490 pounds per cubic foot becomes

$$\rho = \left(490\frac{\text{pounds}}{\text{feet}^3}\right)\left(\frac{1}{2.2}\frac{\text{kilogram}}{\text{pounds}}\right)\left(\frac{39.37}{12}\right)^3\frac{\text{feet}^3}{\text{meters}^3}$$

$$= 7{,}865\frac{\text{kilograms}}{\text{meters}^3}.$$

To convert Rayleigh's breaking stress of $90{,}000\frac{\text{pounds}}{\text{inch}^2}$ to MKS, we want to get the equivalent value in units of $\frac{\text{newtons}}{\text{meter}^2}$. From Newton's second law of motion, $F = ma$, we know 1 newton is the force needed to accelerate a 1-kilogram mass at $1\frac{\text{meter}}{\text{second}^2}$. Since the acceleration of gravity at the Earth's surface is $9.8\frac{\text{meters}}{\text{second}^2}$, we know the force (weight) indicated by our scale would be 9.8 newtons for 2.2 pounds. So, for a 1-pound force we have $9.8/2.2 = 4.45$ newtons. Thus,

$$1\frac{\text{pound}}{\text{inch}^2} = \frac{4.45\,\text{newtons}}{\left(\frac{1}{39.37}\text{meters}\right)^2} = 4.45\left(39.37\right)^2\frac{\text{newtons}}{\text{meter}^2}$$

$$= 6{,}897\frac{\text{newtons}}{\text{meter}^2}.$$

The breaking stress of Rayleigh's steel in MKS units is therefore

$$\sigma = (90{,}000)\left(6{,}897\frac{\text{newtons}}{\text{meter}^2}\right) = 6.2 \times 10^8\frac{\text{newtons}}{\text{meter}^2}.$$

The unit $\frac{newtons}{meter^2}$ has been given a special MKS name of its own: the *pascal* (Pa), in honor of the French scientist Blaise Pascal (1623–1662). So, an alternative MKS measure of the breaking strength of Rayleigh's steel is 620 MPa (megapascals) in the MKS system. Looking back at our result for v, the maximum tangential speed before the ring disintegrates, we should now ask ourselves, does $\sqrt{\frac{\sigma}{\rho}}$ have the units of $\frac{meters}{second}$?

The units of $\sqrt{\frac{\sigma}{\rho}}$ are

$$\sqrt{\frac{newtons/meter^2}{kilograms/meter^3}} = \sqrt{\frac{newtons \times meter}{kilograms}}.$$

Then, remembering that

$$1 \text{ newton} = 1 \text{ kilogram} \times 1\frac{meter}{second^2},$$

we see that

$$\sqrt{\frac{newtons \times meters}{kilograms}} = \sqrt{\frac{\left(kilograms\frac{meters}{second^2}\right)meters}{kilograms}} = \sqrt{\frac{meters^2}{second^2}}$$
$$= \frac{meters}{second},$$

and so our units are okay. All we have left to do is plug in the numbers. At maximum stress in the ring the tangential speed is

$$v = \sqrt{\frac{6.2 \times 10^8}{7,865}\frac{meters}{second}} = 281 \text{ meters per second.}$$

Or, in English units, the maximum tangential speed of the rotating ring is 922 feet per second (which is pretty nearly equal to 628 miles per hour). Or, to put it yet another way, suppose we have a ring with a 1-meter radius (and so a circumference of 2π meters). At maximum stress this ring would be rotating $\frac{281}{2\pi} = 44.7$ times per second. That is, at 2,682 rpm. You definitely don't want to be standing next to this particular Rayleigh ring at 2,683 rpm![5]

In his 1970 novel *Ringworld*, science fiction writer Larry Niven took the concept of a Rayleigh ring—although he never mentioned Rayleigh anywhere—to literally out-of-this-world extremes. There he imagined an artificially constructed world in the shape of a ring spinning around a star. The ring is a band not quite 1 million miles wide and 1,000 meters thick, with a radius of 93 million miles (with the star, of course, at the center). The ring has a tangential speed of 770 miles per second, which gives an outward centrifugal acceleration of 1 g. That is, for beings on the inner face of the band, facing the star, the apparent surface "gravity" would be such that it would be just like walking on Earth.[6]

It's a wonderfully romantic notion, but it does have some problems, of which Niven (who has an undergraduate degree in math) was aware. As he has one of his story characters exclaim, "Consider the tensile strength needed to prevent the structure from disintegrating!" Niven doesn't provide readers with anything more than that, but we can simply reverse our calculations and compute the required strength as

$$\sigma = \rho v^2,$$

[5]In the 1890s this very concern occurred when Rayleigh's friend the physicist Oliver Lodge (1851–1940) conducted experiments in a study of the ether (the mythical substance Victorian physicists at one time believed filled all of what appeared to be empty space, through which electromagnetic waves could travel). Lodge's experimental setup involved the high-speed rotation of massive steel plates, and by Christmas 1891 he was running his machine at 2,800 rpm. At that speed one of Lodge's friends worried that if there were any flaws in the plates they could disintegrate, and, as Lodge himself wrote in a lab notebook, "we should have our heads cut off." For more on this fascinating but little-known episode in physics, see Bruce Hunt, "Experimenting on the Ether: Oliver J. Lodge and the Great Whirling Machine," *Historical Studies in the Physical and Biological Sciences*, 16 (no. 1) 1986, pp. 111–134.

[6]With $v = 770 \times 5,280$ feet/second $= 4.07 \times 10^6$ feet/second, and $R = 93,000,000 \times 5,280$ feet $= 4.91 \times 10^{11}$ feet, the acceleration is $\frac{v^2}{R} = \frac{(4.07 \times 10^6)^2}{4.91 \times 10^{11}}$ $\frac{\text{feet}}{\text{second}^2} = 33.7 \frac{\text{feet}}{\text{second}^2}$, which is, indeed, just a bit over 1 g.

where $v = \frac{4.07 \times 10^6 \text{ feet/second}}{3.28 \text{ feet/meter}} = 1.24 \times 10^6 \text{meters/second}$. To calculate Ringworld's ρ, I'll divide its mass (Niven says it is "two times ten to the thirtieth power" in grams[7]), divided by the volume of the ring. The volume of Ringworld is its circumference times its width times its thickness, and so is

$$\left(2\pi \frac{4.91 \times 10^{11}}{3.28} \text{meters}\right) \times \left(\frac{10^6 \times 5{,}280}{3.28} \text{meters}\right) \times 10^3 \text{meters}$$

$$= 1.5 \times 10^{24} \text{meters}^3.$$

Thus,

$$\rho = \frac{2 \times 10^{27} \text{ kilograms}}{1.5 \times 10^{24} \text{ meters}^3} = 1.33 \times 10^3 \frac{\text{kilograms}}{\text{meters}^3},$$

which is, perhaps surprisingly, *less* than the density of Rayleigh's steel.

But even a Ringworld made of Rayleigh steel couldn't stay together, because

$$\sigma = 1.33 \times 10^3 \frac{\text{kilograms}}{\text{meters}^3} \times \left(1.24 \times 10^6 \frac{\text{meters}}{\text{second}}\right)^2 = 2 \times 10^{15} \frac{\text{newtons}}{\text{meter}^2},$$

which is more than *3 million times greater* than that of Rayleigh's steel.

As a final note, as I mentioned earlier, the metric system encountered not just a little resistance when introduced in England, resistance spoofed in a funny 1864 poem ("The Three-Foot Rule") by the Scottish engineer and physicist William Rankine (1820–1872), in what appears to be a plea to favor the 3-foot yard over the newfangled, slightly longer meter as a unit of length:

> When I was a bound apprentice, and learned to use my hands,
> Folk never talked of measures, that came from foreign lands;
> Now I'm a British Workman, too old to go to school,
> So whether the chisel or file I hold, I'll stick to my three-foot rule.
> Some talk of millimeters, and some of kilogrammes,

---

[7]This is essentially the mass of Jupiter: 2×10^{27} kilograms.

And some of deciliters, to measure beer and drams;
Now I'm a British Workman, too old to go to school,
So by pounds I'll eat, and by quarts I'll drink and
I'll work by my three-foot rule.

The similarity of those lines to the lyrics of a Gilbert & Sullivan comic opera is hard to miss!

 *Acknowledgments*

Many of the people I've worked with in the past are back again for this book, behind the scenes but all so very important. They include Vickie Kearn, my always supportive Princeton editor; her efficient assistant, Lauren Bucca; the book's terrific production editor, Deborah Tegarden; the Press's talented artists Dimitri Karetnikov and Carmina Alvarez-Gaffin, who somehow turned my crude illustration scrawls into art; the book's laser-focused copy editor, Barbara Liguori, who saved me (more than once) from embarrassing errors; and the anonymous reviewers of my original proposal for this book, whose critical comments were most helpful as I wrote. Loma Karklins, at the Caltech Archives, was superhelpful in my quest for the frontispiece photograph of Feynman's last blackboard (and since I myself am a graduate of Caltech, Loma was pleased to give me a discounted fee for the use of the photograph!). My University of New Hampshire administrative assistant, Mrs. Kathy Reynolds, was enormously helpful in the production of the final typescript submitted to Princeton editorial. A version of Appendix 4 ("Rayleigh's Rotating Ring") appeared earlier in *The Mathematical Intelligencer* (Summer 2017), and I thank Bob Burckel and Larry Weaver (mathematician and physicist, respectively, at Kansas State University) for pushing me to make that discussion better than in my first attempt.

Special thanks are due to the ever-pleasant staff of Me & Ollie's Bakery, Bread and Café, on Water Street in Exeter, New Hampshire. As I sat almost daily in my cozy little nook by a window, next to the town bandstand, surrounded by happily chattering Phillips Exeter Academy high school students from just up the street (all of whom carefully avoided making eye contact with the strange old guy mysteriously scribbling away on papers scattered all over the table), and energized by endless cups of hot pumpkin spice coffee (particularly during February snowstorms) to keep me awake, the mathematical physics seemed to just flow off my pen.

The exotic cat that first appeared on the cover of *In Praise of Simple Physics* (Princeton, 2016) is back again for this book, and I wish to express my special thanks to the artist, Anne Karetnikov, who created her. The expression on that cat's face catches, *perfectly*, the spirit in which I wrote this book.

And finally, I thank my wife, Patricia Ann, who has—now for more than 55 years—put up with my world-class talent for raising the entropy of our home at an absurd rate. With only minor grumbling (well, maybe not *always* so minor), she has allowed me to occupy the mess I routinely make of the local surroundings. Perhaps she has simply given up trying to change me, but I prefer to think it's because she loves me. I know I love her.

February 2018

Index

Also by Paul J. Nahin

Computing Nearly 200 Perplexing Definite Integrals from Physics, Engineering, and Mathematics (Plus 60 Challenge Problems with Complete, Detailed Solutions)

In Praise of Simple Physics: *The Science and Mathematics behind Everyday Questions*

Time Machine Tales: *The Science Fiction Adventures and Philosophical Puzzles of Time Travel*

Transients for Electrical Engineers: *Elementary Switched Circuit Analysis in the Time and Laplace Transform Domains (with a Touch of MATLAB®)*